Communications
in Computer and Information Science   2094

Rationale

The CCIS series is devoted to the publication of proceedings of computer science conferences. Its aim is to efficiently disseminate original research results in informatics in printed and electronic form. While the focus is on publication of peer-reviewed full papers presenting mature work, inclusion of reviewed short papers reporting on work in progress is welcome, too. Besides globally relevant meetings with internationally representative program committees guaranteeing a strict peer-reviewing and paper selection process, conferences run by societies or of high regional or national relevance are also considered for publication.

Topics

The topical scope of CCIS spans the entire spectrum of informatics ranging from foundational topics in the theory of computing to information and communications science and technology and a broad variety of interdisciplinary application fields.

Information for Volume Editors and Authors

Publication in CCIS is free of charge. No royalties are paid, however, we offer registered conference participants temporary free access to the online version of the conference proceedings on SpringerLink (http://link.springer.com) by means of an http referrer from the conference website and/or a number of complimentary printed copies, as specified in the official acceptance email of the event.

CCIS proceedings can be published in time for distribution at conferences or as post-proceedings, and delivered in the form of printed books and/or electronically as USBs and/or e-content licenses for accessing proceedings at SpringerLink. Furthermore, CCIS proceedings are included in the CCIS electronic book series hosted in the SpringerLink digital library at http://link.springer.com/bookseries/7899. Conferences publishing in CCIS are allowed to use Online Conference Service (OCS) for managing the whole proceedings lifecycle (from submission and reviewing to preparing for publication) free of charge.

Publication process

The language of publication is exclusively English. Authors publishing in CCIS have to sign the Springer CCIS copyright transfer form, however, they are free to use their material published in CCIS for substantially changed, more elaborate subsequent publications elsewhere. For the preparation of the camera-ready papers/files, authors have to strictly adhere to the Springer CCIS Authors' Instructions and are strongly encouraged to use the CCIS LaTeX style files or templates.

Abstracting/Indexing

CCIS is abstracted/indexed in DBLP, Google Scholar, EI-Compendex, Mathematical Reviews, SCImago, Scopus. CCIS volumes are also submitted for the inclusion in ISI Proceedings.

How to start

To start the evaluation of your proposal for inclusion in the CCIS series, please send an e-mail to ccis@springer.com.

Xiangyu Song · Ruyi Feng · Yunliang Chen ·
Jianxin Li · Geyong Min
Editors

Web and Big Data

APWeb-WAIM 2023 International Workshops

KGMA 2023 and SemiBDMA 2023
Wuhan, China, October 6–8, 2023
Proceedings

 Springer

Editors
Xiangyu Song 🆔
Peng Cheng Laboratory
Shenzhen, China

Ruyi Feng 🆔
China University of Geosciences
Wuhan, China

Yunliang Chen 🆔
China University of Geosciences
Wuhan, China

Jianxin Li 🆔
Deakin University
Burwood, VIC, Australia

Geyong Min 🆔
University of Exeter
Exeter, UK

ISSN 1865-0929   ISSN 1865-0937 (electronic)
Communications in Computer and Information Science
ISBN 978-981-97-2990-6   ISBN 978-981-97-2991-3 (eBook)
https://doi.org/10.1007/978-981-97-2991-3

This Springer imprint is published by the registered company Springer Nature Singapore Pte Ltd.
The registered company address is: 152 Beach Road, #21-01/04 Gateway East, Singapore 189721, Singapore

If disposing of this product, please recycle the paper.

Preface

The Asia-Pacific Web (APWeb) and Web-Age Information Management (WAIM) Joint International Conference on Web and Big Data (APWeb-WAIM) is a leading international conference for researchers, practitioners, developers, and users to share and exchange their cutting-edge ideas, results, experiences, techniques, and applications in connection with all aspects of web and big data management.

The conference invites original research papers on the theory, design, and implementation of data management systems. As the 7th edition in the increasingly popular series, APWeb-WAIM 2023 was held in Wuhan, China, 6–8 October 2023. Along with the main conference, the APWeb-WAIM workshops provide an international forum for researchers to discuss and share their pioneering works. This APWeb-WAIM 2023 workshops volume contains the papers accepted for the two workshops held in conjunction with APWeb-WAIM 2023. The two workshops were selected after a public call-for-proposals process, and each of them had a focus on a specific area that contributed to the main themes of the APWeb-WAIM conference. After the single-blinded review process, out of 15 submissions, the two workshops accepted a total of 7 papers, marking an acceptance rate of 46.67%. Each submission was peer-reviewed by 3–4 reviewers. The two workshops were as follows:

- The Sixth International Workshop on Knowledge Graph Management and Applications (KGMA 2023)
- The Fifth International Workshop on Semi-structured Big Data Management and Applications (SemiBDMA 2023)

As a joint effort, all organizers of APWeb-WAIM conferences and workshops, including this and previous editions, have made APWeb-WAIM a valuable trademark through their work. We would like to express our thanks to all the workshop organizers and Program Committee members for their great efforts in making the APWeb-WAIM 2023 workshops such a great success. Last but not least, we are grateful to the main conference organizers for their leadership and generous support, without which this APWeb-WAIM 2023 workshop volume would not have been possible.

December 2023

Yunliang Chen
Jianxin Li
Geyong Min
David A. Yuen
Ruyi Feng
Xiangyu Song

Organization

APWeb-WAIM 2023 Proceedings Co-chairs

Ruyi Feng   China University of Geosciences, China
Xiangyu Song   Swinburne University of Technology, Australia

SemiBDMA 2023

Workshop Co-chairs

Baoyan Song   Liaoning University, China
Xiaoguang Li   Liaoning University, China
Linlin Ding   Liaoning University, China
Yuefeng Du   Liaoning University, China

Program Committee Members

Ye Yuan   Northeastern University, China
Xiangmin Zhou   RMIT University, Australia
Jianxin Li   Swinburne University of Technology, Australia
Bo Ning   Dalian Maritime University, China
Yongjiao Sun   Northeastern University, China
Yulei Fan   Zhejiang University of Technology, China
Guohui Ding   Shenyang Aerospace University, China
Bo Lu   Dalian Nationalities University, China
Linlin Ding   Liaoning University, China
Xiaohuan Shan   Liaoning University, China
Yuefeng Du   Liaoning University, China
Tingting Liu   Liaoning University, China

KGMA 2023

Workshop Co-chairs

Xiang Zhao   National University of Defense Technology,
                                   China
Xin Wang   Tianjin University, China

Program Committee Members

Huajun Chen   Zhejiang University, China
Wei Hu   Nanjing University, China
Saiful Islam   Griffith University, Australia
Jiaheng Lu   University of Helsinki, Finland
Jianxin Li   Deakin University, Australia
Ronghua Li   Beijing Institute of Technology, China
Jeff Z. Pan   University of Aberdeen, UK
Jijun Tang   University of South Carolina, USA
Haofen Wang   Tongji University, China
Hongzhi Wang   Harbin Institute of Technology, China
Junhu Wang   Griffith University, Australia
Meng Wang   Southeast University, China
Xiaoling Wang   East China Normal University, China
Xuguang Ren   G42 Inception Institute of Artificial Intelligence,
                                   UAE
Guohui Xiao   Free University of Bozen-Bolzano, Italy
Zhuoming Xu   Hohai University, China
Qingpeng Zhang   City University of Hong Kong, China
Xiaowang Zhang   Tianjin University, China
Weiren Yu   University of Warwick, UK

Contents

KGMA 2023

A Bidirectional Question-Answering System using Large Language Models and Knowledge Graphs

Lifan Han, Xin Wang[✉], Zhao Li, Heyi Zhang, and Zirui Chen

College of Intelligence and Computing, Tianjin University, Tianjin, China
{hanlf,wangx,lizh,zhy111,zrchen}@tju.edu.cn

Abstract. The integration of Large Language Models (LLMs) and Knowledge Graphs (KGs) has emerged as a vibrant research area in the field of Natural Language Processing (NLP). However, existing approaches need help effectively harnessing the complementary strengths of LLMs and KGs. In this paper, we propose a novel system that addresses this gap by enabling bidirectional conversion between LLMs and KGs. We leverage external knowledge to enhance LLMs for domain-specific responses and fine-tune LLMs for information extraction to construct the Knowledge Graph. Moreover, users can interact with the KG, initiating new rounds of questioning in LLMs. The evaluation results highlight the effectiveness of our approach. Our system showcases the potential of combining LLMs and KGs, paving the way for advanced natural language understanding and generation in various domains.

Keywords: Large Language Model · Knowledge Graph · Natural Language Processing

1 Introduction

In recent years, the emergence of Large Language Models (LLMs) such as Chat-GPT[1] has garnered significant attention in Natural Language Processing (NLP). These models, trained on large-scale corpora, have demonstrated remarkable capabilities in processing and interpreting natural language and understanding human communication. However, the success of LLMs has primarily been focused on statistically-driven language generation, while challenges persist in achieving precise semantic understanding and accurate text output [1].

Simultaneously, Knowledge Graphs (KGs) have gained significance as a structured representation of knowledge. KGs encompass factual information that can provide LLMs with prior and background knowledge, thereby enhancing their performance in semantic understanding and generation tasks [17]. However, constructing KGs is challenging, and the advent of LLMs offers a potential solution to address this challenge.

Hence, integrating LLMs and KGs, leveraging their respective strengths, is a natural progression. In recent years, numerous studies have explored the combination of LLMs and KGs [10]. However, a systematic framework for bidirectional

---

[1] https://openai.com/blog/chatgpt

© The Author(s), under exclusive license to Springer Nature Singapore Pte Ltd. 2024
X. Song et al. (Eds.): APWeb-WAIM 2023 Workshops, CCIS 2094, pp. 3–10, 2024.
https://doi.org/10.1007/978-981-97-2991-3_1

connectivity between LLMs and KGs must be improved. Bidirectional connectivity refers to the information exchange and knowledge enhancement from LLMs to KGs and vice versa.

The main contributions of this paper are as follows:

– We propose a novel system that achieves bidirectional connectivity between LLMs and KGs, enabling the construction and updating of KGs based on questions generated by LLMs while enhancing LLMs' performance with knowledge from KGs.
– We design a click-based interaction approach that allows users to conveniently select entities or relations of interest from KGs and use them as input questions for LLMs.
– We conducted experiments on both subjective and objective questions, providing empirical evidence for the effectiveness and superiority of our system in the context of question-answering (QA) tasks. Furthermore, we curated datasets and conducted experiments in the field of information extraction, investigating model performance under various conditions.

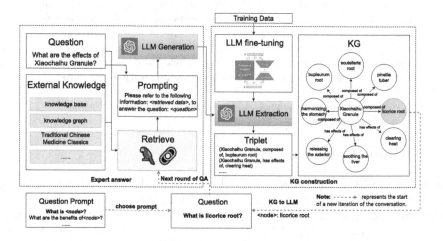

Fig. 1. The overall architecture of our system.

2   System Architecture

The Figure 1 overviews the main architecture of the system. The system architecture comprises two primary components: Expert Answer and KG Construction. Subsequent sections will provide detailed explanations of each component.

Expert Answer. We have incorporated external knowledge as a knowledge repository to ensure a more professional output from our system. When given a user query, we leverage LangChain[2] to retrieve relevant professional expertise

---

[2] https://www.langchain.com.

from this repository. Combining the retrieved professional knowledge with the user's question creates a comprehensive prompt inputted into LLM. This process enables us to obtain answers that are linguistically accurate and infused with specialized expertise. By incorporating domain-specific information directly into the prompt, we ensure that the system's responses are imbued with professionalism and domain expertise.

KG Construction. To construct the KG, we developed an information extraction pipeline. Initially, we built an annotated dataset for information extraction by manually annotating triplets within sentences. Subsequently, we fine-tuned the LLM using this annotated dataset to enable it to learn patterns and structures required for information extraction. The trained LLM can generate triplets containing entities and relationships from input text. Using these triplets, we visually structured the information and displayed it in an interface, forming a KG. This process can be understood as the transformation from LLM to KG.

Furthermore, we established bidirectional connectivity between the KG and the LLM. By selecting a specific node within the KG, such as in Figure 1, clicking on "licorice root" and choosing "benefits," we initiate a new round of questioning in the LLM, forming a prompt like "What are the benefits of licorice root?" This process represents the transformation from KG to LLM. This interactive process allows users to explore and retrieve information from the KG through a natural and intuitive interface.

Through this pipeline, users can engage in an interactive exploration of the KG, leveraging the bidirectional interaction between the LLM and the KG. This system enables an integration of the language processing capabilities of the LLM with the structured knowledge representation offered by the KG, empowering users to navigate and retrieve information efficiently.

3   Demonstration and Evaluation

3.1   Web Interface

As depicted in Figure 2, the system interface provides two demonstrations: one showcasing the flow from LLM to KG and the other demonstrating the interaction from KG to LLM. In the LLM to KG demonstration, users input their questions, click the "Send" button, and then wait for the LLM to generate the answer and perform information extraction. Subsequently, they can obtain the solution and a visual representation of the KG. On the other hand, in the KG to LLM interaction, users can right-click on a specific node, triggering a pop-up panel where they can select the desired relationship to inquire about. This selection serves as the question for a new round of interaction between the user and the LLM, resulting in an updated answer and the generation of an updated KG.

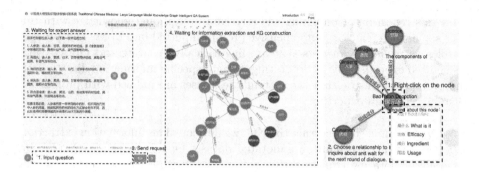

Fig. 2. Screenshots of Our System, with the Left Image Demonstrating LLM to KG and the Right Image Demonstrating KG to LLM.

Table 1. Results of the objective Evaluation. Accuracy values are represented. SQ, MQ, DQ, and ALL denote the average accuracy for Simple Questions, Medium Questions, Difficult Questions, and All questions, respectively.

Model	SQ	MQ	DQ	ALL
ChatGLM + LangChain	**0.85**	0.65	0.55	0.70
ChatGLM[2]	0.75	0.55	0.30	0.58
ChatGPT 1	**0.85**	**0.70**	**0.60**	**0.74**

Table 2. Comparison of subjective evaluation results by three TCM experts for different models. The numbers in the table represent the count of questions for which each model was deemed to provide the best answers by the experts.

Model	Expert 1	Expert 2	Expert 3	Avg. Count
ChatGLM + LangChain	**37**	**42**	**42**	**40.3**
ChatGLM [2]	27	21	26	24.7
ChatGPT 1	36	37	32	35

3.2 Evaluation

In this section, we conducted an evaluation within the domain of Traditional Chinese Medicine (TCM), encompassing both subjective and objective assessments of the Question-Answering (QA) system's performance, as well as an evaluation of the LLM fine-tuned for information extraction. It is worth noting that all experiments in this section were carried out in the MindSpore 1.10.1 environment, leveraging Ascend910 NPU hardware for computational acceleration.

Subjective Evaluation. To assess the system's performance, we curated a set of 100 questions relevant to TCM and inputted them into the three models to generate answers. Subsequently, we invited three TCM experts to evaluate the content generated by each model and recorded their assessments of which model produced the most satisfactory answers for each question. The evaluation results, presented in Table 2, indicate the average number of questions for which each model was considered to provide the most satisfactory answers by the three experts.

Objective Evaluation. In the realm of objective evaluation, we employed a meticulous approach to assess model performance at different levels of difficulty. We collected a set of fifty multiple-choice questions within the domain of TCM. These questions spanned varying degrees of complexity. Subsequently, we employed a prompt-based approach, inviting LLMs to provide responses. The accuracy of each model's responses was meticulously computed for each level of question difficulty. The comprehensive results are presented in Table 1.

Table 3. Information extraction performance evaluation metrics. Precision, Recall, and F1 Score are reported for different model configurations.

	Precision	Recall	F1
Origin	0	0	0
+ zero-shot	0.11	0.05	0.07
+ one-shot	0.27	0.14	0.19
+ five-shot	0.42	0.25	0.31
+ P-tuning v2	**0.79**	**0.68**	**0.73**

Information Extraction Evaluation. In the context of information extraction, we conducted model fine-tuning utilizing the P-tuning v2 [7], configuring key hyperparameters as follows: a learning rate of 2e-2, a batch size of 2, and training over 2 epochs. Each experiment was repeated 5 times, and the performance metrics were averaged to ensure robustness and reduce the impact of random variations. Our dataset comprises 100 meticulously curated collections of TCM prescriptions. It's important to note that this dataset consists of manually annotated triplets extracted from TCM prescription data and has undergone rigorous validation for accuracy and reliability. These triplets are meticulously designed to encompass a wide spectrum of TCM knowledge, including four categories of entities: prescription formulations, medicinal herbs, ailments, and medicinal herb properties. Additionally, the dataset includes seven types of relationships, covering constituent components, corresponding ailments, contraindications, and more. To substantiate the effectiveness of our information extraction model, we conducted a comprehensive evaluation. The primary aim of this evaluation was to extract all TCM-related triplets from TCM texts and meticulously compare the extracted results with triplets manually annotated by domain experts. We conducted a comparative analysis of the performance between the original Chat-GLM model and various modifications, including zero-shot, one-shot, five-shot, and P-tuning v2 fine-tuning. The summarized results are presented in Table 3.

4   Related Work

Enhancing KGs with LLMs. Recent research has introduced several approaches for enhancing KGs using LLMs. In the realm of KG completion, KG-BERT [20] modifies the original BERT [3] architecture by representing inputs as entity and relation triples and computes a scoring function for these triples, enabling Knowledge Graph Completion (KGC). KGT5 [14] formulates both knowledge graph link prediction and question answering as sequence-to-sequence (seq2seq) tasks. A straightforward encoder-decoder Transformer is trained on these tasks, with regularization using link prediction objectives during question answering training. Information extraction represents a fundamental and pivotal task in the construction of knowledge graphs. It encompasses tasks such as extracting entity-relation triples (RE), recognizing named entities (NER), and extracting events (EE) [4–6,16]. Handling substantial labeled data is consistently complex and time-consuming. Recent research has also proposed the use of LLMs to undertake these tasks. GenerativeNER [19] leverages a sequence-to-sequence LLM with a pointer mechanism to generate entity sequences, effectively addressing NER tasks. LSR [9] introduces a graph-based approach that incorporates graph structures on top of LLMs to better extract relationships within KGs. Grapher [8] tackles the challenge of constructing KGs from text, presenting an innovative end-to-end, multi-stage KG construction system. It begins by utilizing an LLM to generate KG entities and efficiently extracts KG information from textual descriptions through the use of a simple edge construction head.

Enhancing LLMs with KGs. LLMs are renowned for their state-of-the-art performance across various NLP tasks. However, they often suffer from the problem of hallucinations and exhibit suboptimal performance in scenarios where knowledge traceability, timeliness, and accuracy are crucial [11]. Given that KGs store a vast amount of knowledge in a structured manner, they offer a valuable resource for LLMs to acquire knowledge. Some researchers have proposed the integration of KGs into LLMs during the pretraining phase [13,18,21]. Furthermore, frameworks have been introduced to leverage KGs to enhance the depth and responsible reasoning capabilities of LLMs. Think-on-Graph (ToG) [15] is one such framework that can identify entities relevant to a given question, engage in exploration and inference, and retrieve pertinent triples from external knowledge databases.

5   Conclusion

Our system introduces a groundbreaking approach that enables seamless interaction between LLMs and KGs. Through bidirectional connectivity, our system leverages LLMs to enhance KG construction while integrating KG knowledge for improved LLM performance. Both subjective and objective evaluations, as well as the information extraction performance analysis, collectively attest to the effectiveness of our approach. These assessments collectively demonstrate the system's potential to advance natural language understanding and generation across various domains.

Acknowledgement. This work is supported by the CAAI-Huawei MindSpore Open Fund (2022037A).

References

1. Dong, C., et al.: A survey of natural language generation. ACM Comput. Surv. **55**(8), 1–38 (2022)
2. Du, Z., et al.: GLM: general language model pretraining with autoregressive blank infilling. In: Proceedings of the 60th Annual Meeting of the Association for Computational Linguistics, Vol. 1, pp. 320–335 (2022)
3. Kenton, J.D.M.W.C., Toutanova, L.K.: BERT: pre-training of deep bidirectional transformers for language understanding. In: Proceedings of NAACL-HLT, pp. 4171–4186 (2019)
4. Kumar, S.: A survey of deep learning methods for relation extraction. arXiv preprint. arXiv:1705.03645 (2017)
5. Li, J., Sun, A., Han, J., Li, C.: A survey on deep learning for named entity recognition. IEEE Trans. Knowl. Data Eng. **34**(1), 50–70 (2020)
6. Liu, J., Chen, Y., Liu, K., Bi, W., Liu, X.: Event extraction as machine reading comprehension. In: Proceedings of the 2020 conference on empirical methods in natural language processing (EMNLP), pp. 1641–1651 (2020)
7. Liu, X., et al.: P-tuning v2: prompt tuning can be comparable to fine-tuning universally across scales and tasks. arXiv preprint. arXiv:2110.07602 (2021)
8. Melnyk, I., Dognin, P., Das, P.: Grapher: multi-stage knowledge graph construction using pretrained language models. In: NeurIPS 2021 Workshop on Deep Generative Models and Downstream Applications (2021)
9. Nan, G., Guo, Z., Sekulić, I., Lu, W.: Reasoning with latent structure refinement for document-level relation extraction. In: Proceedings of the 58th Annual Meeting of the Association for Computational Linguistics, pp. 1546–1557 (2020)
10. Pan, S., Luo, L., Wang, Y., Chen, C., Wang, J., Wu, X.: Unifying large language models and knowledge graphs: a roadmap. arXiv preprint arXiv:2306.08302 (2023)
11. Petroni, F., et al.: Kilt: a benchmark for knowledge intensive language tasks. In: Proceedings of the 2021 Conference of the North American Chapter of the Association for Computational Linguistics: Human Language Technologies, pp. 2523–2544 (2021)
12. Radford, A., Narasimhan, K., Salimans, T., Sutskever, I.: Improving language understanding by generative pre-training
13. Rosset, C., Xiong, C., Phan, M., Song, X., Bennett, P., Tiwary, S.: Knowledge-aware language model pretraining. arXiv preprint. arXiv:2007.00655 (2020)
14. Saxena, A., Kochsiek, A., Gemulla, R.: Sequence-to-sequence knowledge graph completion and question answering. In: Proceedings of the 60th Annual Meeting of the Association for Computational Linguistics (Vol 1: Long Papers), pp. 2814–2828 (2022)
15. Sun, J., et al.: Think-on-graph: Deep and responsible reasoning of large language model with knowledge graph. arXiv preprint. arXiv:2307.07697 (2023)

16. Tjong Kim Sang, E.F., De Meulder, F.: Introduction to the CoNLL-2003 shared task: Language-independent named entity recognition. In: Proceedings of the Seventh Conference on Natural Language Learning at HLT-NAACL 2003, pp. 142–147 (2003), https://aclanthology.org/W03-0419

17. Wang, X., et al.: Improving natural language inference using external knowledge in the science questions domain. In: Proceedings of the AAAI Conference on Artificial Intelligence. vol. 33, pp. 7208–7215 (2019)

18. Wang, X., et al.: Kepler: a unified model for knowledge embedding and pre-trained language representation. Trans. Assoc. Comput. Linguist. **9**, 176–194 (2021)

19. Yan, H., Gui, T., Dai, J., Guo, Q., Zhang, Z., Qiu, X.: A unified generative framework for various NER subtasks. In: Proceedings of the 59th Annual Meeting of the Association for Computational Linguistics and the 11th International Joint Conference on Natural Language Processing (Vol 1: Long Papers), pp. 5808–5822 (2021)

20. Yao, L., Mao, C., Luo, Y.: KG-BERT: Bert for knowledge graph completion. arXiv preprint. arXiv:1909.03193 (2019)

21. Zhang, Z., Han, X., Liu, Z., Jiang, X., Sun, M., Liu, Q.: Ernie: Enhanced language representation with informative entities. In: Proceedings of the 57th Annual Meeting of the Association for Computational Linguistics, pp. 1441–1451 (2019)

A Comprehensive Review of Relation Prediction Techniques in Knowledge Graph

Yuxuan Lu, Shiyu Yang[(✉)], and Benzhao Tang

Guangzhou University, Guangzhou, China
{powerlyx,benzhaotang}@e.gzhu.edu.cn, syyang@gzhu.edu.cn

Abstract. Knowledge graphs organize entity relations using a graph structure, facilitating knowledge representation. In research, relation prediction within knowledge graphs plays a crucial role, aiding inference, latent knowledge discovery, and revealing intricate associations between entities. We present an overview of this field's development and methods. Initially, we introduce fundamental concepts, relation prediction task definitions, and evaluation metrics. Subsequently, we delve into research, spanning rule-based, statistical, and modern approaches like representation learning, deep learning and large language models. We explore transductive and inductive learning modes, discussing their relevance in relation prediction, and classify and summarize these methods. Additionally, we evaluate method strengths, weaknesses, and suitable scenarios, providing insights. Finally, we address future research directions and challenges in knowledge graph relation prediction, offering guidance for further study and practical applications.

Keywords: Knowledge Graph · Relation Prediction · Large Language Model

1 Introduction

In 2012, Google introduced the concept of knowledge graphs to enhance its search engine capabilities. A knowledge graph(KG) is essentially a vast repository of entities and their relations, designed to represent semantic information about the real world. Some common knowledge graph such as Freebase [3], YAGO [44], DBpedia [2], and Wikidata [49] has garnered substantial attention in the academic community, especially with the rapid advancements in natural language processing (NLP) technology. Knowledge graphs store facts as triples, specifying relations between entities. This triples hold rich information and are structured as graphs, allowing for direct application of graph algorithms. Consequently, it is extensively used in artificial intelligence applications like question answering [64], relation extraction, text generation [3], and recommendation systems. However, as knowledge graph data scales up, creating large-scale graphs demands significant time and effort. Incomplete knowledge graphs can hamper downstream

X. Song et al. (Eds.): APWeb-WAIM 2023 Workshops, CCIS 2094, pp. 11–24, 2024.
https://doi.org/10.1007/978-981-97-2991-3_2

applications, prompting the need for relation prediction tasks. This task aims to mine missing triples, automatically uncovering hidden relations and enhancing the graph's completeness. Yet, relation prediction research faces challenges. Heterogeneous information within knowledge graphs, spanning different types, scales, and domains, poses a significant hurdle. Additionally, complex semantic attributes in relations, such as one-to-many or hierarchical connections, can be challenging to understand and predict.

Based on the nature of prediction targets, these tasks can be categorized into two types: transductive and inductive, depending on whether they can predict entities and relations that have not appeared in the training process. Transductive setting involves forecasting entities and relations that have appeared in the training process, assessing the existence of a particular relation under the present conditions of known entities and relations. However, in reality, new entities and relations continually emerge, and the cost of retraining models is substantial. Consequently, adapting to dynamically evolving knowledge graphs becomes pivotal. Hence, inductive relation prediction tasks effectively function as logical inference problems. They involve the prediction of entities and relations that haven't been seen during training. This can achieved by mining probabilistic logical rules within the knowledge graph, employing these rules to infer the existence or absence of certain relations.

In this paper, we provide a comprehensive review of currently available KG relation prediction techniques, including use rules, KG embeddings, as well as those that further leverage graph neural network and large language models (LLMs). There are several survey papers about both KG embeddings and relation prediction. Wang et al. [52] focus on KG embeddings. Furthermore, Wang et al. [50] focus on knowledge graph embedding techniques in relation prediction. Unlike these papers, our review has a more specific topic, focusing on logics and inductive settings. The survey also introduces the latest research on large language models.

2   Related Works

In this section, we review and discuss prior review papers related to our research, which provide a comprehensive understanding of the field and valuable background information for our study. Wang et al. [52] provided a comprehensive overview of knowledge graph embedding methods, introducing both traditional embedding approaches and neural network approaches, offering us contextual information to comprehend the evolution of embedding techniques. Wang et al. [50] concentrated on the utilization of knowledge graph embedding techniques in link prediction tasks. It underscored the significance of embedding methods within knowledge graphs and provided a comprehensive introduction to evaluation methods for relation prediction. This has contributed useful methodological background to our research. Siddhant Arora [1] summarized the application of graph neural networks in knowledge graph completion, elaborating on the advantages, disadvantages, and application scenarios among various models of graph neural networks. This has provided valuable insights into graph neural network methods for our research.

While these review papers offer profound insights, our unique contribution lies in exploring the importance of inductive reasoning models in real-world applications. We categorize and classify existing models from both transductive and inductive perspectives. Additionally, we surveyed recent research on large language models in relation prediction, providing a fresh perspective and insights for the field.

3   Methods

We first introduce rule-based methods, representation learning-based methods, and deep learning-based methods for knowledge graph relation prediction tasks. Then, in real-life scenarios, new entities and relations are continually added to knowledge graphs, we can further categorize relation prediction methods into transductive and inductive approaches. Finally, we discuss the KG applications with LLMs.

3.1   Rule-Based Methods

Rule-based method relies on predefined rules and logical relation to discover new relations through logical inference. These rules can be established based on domain knowledge, grammatical structures, or pattern matching, and are then applied to entities and relations within the knowledge graph.

NELL [37] is an automated knowledge graph construction system that integrates methods based on pattern matching, rules, and machine learning techniques. It utilizes first-order logic learning algorithms to infer new relation instances. Richardson et al. [39] introduced the Markov Logic Network (MLN), laying the foundation for research in relation prediction tasks. MLN cleverly combines first-order logic and probabilistic graphical models, assigning learnable weights to each rule to effectively handle uncertainty and complex dependencies in reasoning. However, MLN faces challenges in scalability and requires careful parameter tuning. Subsequently, AMIE [18] emerged in the field of knowledge base mining, focusing on the extraction of logical rules, particularly suitable for open-world assumption (OWA) scenarios, but it performs less effectively on large-scale knowledge bases. To address scalability issues, AMIE+ [17] extends AMIE by refining rules and enhancing confidence assessment, enabling it to handle exceptionally large knowledge bases. Recently, TensorLog [10] integrate logical reasoning with neural network technology, providing a differentiable inductive database system. It not only serves for the representation and reasoning of logical rules but also supports gradient computation for rule-based optimization and learning. Neural LP [58] is based on TensorLog, it introduced an innovative fusion of parameter learning and structural learning for first-order logic rules within a neural network framework, enabling end-to-end learning. The primary advantage of Neural LP lies in its capacity for end-to-end learning, directly acquiring knowledge graph representations and rule parameters from raw data. However, for large-scale knowledge graphs, model training and inference processes are comparatively slower. Building upon Neural LP, DRUM [40] extended rule learning and inference by introducing variable-length patterns, enabling the acquisition of richer rules for more precise inference. The incorporation of an attention mechanism simulates probabilistic logic rules.

Then, Logic Attention Network (LAN) [51] introduced a novel node neighborhood aggregator that combines attention weights based on rules and network structures, particularly suitable for inductive relation prediction in knowledge graphs. In comparison to the AMIE approach, RuleN [32] approximates rule confidence by randomly selecting triplets within the knowledge graph, utilizing a stochastic selection method to mine longer rules, thereby enhancing inference performance.

Although rule-based relation inference methods have the capability to capture implicit information and associations in knowledge graphs, their performance is constrained by the quality and quantity of rules. Firstly, writing and fine-tuning rules typically require the involvement of domain experts, which demands significant human efforts and domain knowledge. Secondly, rule-based methods have limited expressive ability, making it challenging to handle complex reasoning tasks.

3.2   Representation Learning-Based Methods

Representation Learning-based method centers on mapping entities and relations into a continuous, low-dimensional embedding space. By learning vector representations of entities and relations within this space, we can capture and encode complex patterns between entities and relations. We introduced translation-based and semantic matching-based methods as follows.

I. Translation-Based. Translation-based method constitute a significant approach in representation learning. The fundamental concept behind these methods is to treat relations as transformations between entities within an embedding space. The goal is to learn embeddings in such a way that, for any given entity pair (h, t) and relation r, entity h, when 'translated' by relation r, closely approximates entity t. TransE [5] and its variants represent the earliest and most emblematic translation-based approaches. TransE leverages geometric relations in vector space for predictions, enhancing the interpretability of its results. However, it struggles with complex relations $(1 - n, n - 1, n - n)$, excelling primarily in $(1 - n)$ relation predictions. Currently, there are several variations of TransE model aimed at addressing its limitations. TransH [53] introduces the concept of hyperplanes by assigning a hyperplane to each relation, effectively resolving TransE's issues with handling complex relation. TransR [29], on the other hand, assumes that each relation has an independent vector space and employs projection matrices to map entities from the entity space to the relation space. Figure 1 illustrates the entity and relation space of TransE, R, H models. TransD

[24] extends TransR by introducing dynamic mapping matrices to reduce model parameters while maintaining flexibility. However, the TransD model still faces challenges such as sensitivity to hyperparameters and computational complexity. TransA [55] proposes an adaptive metric learning method to address the oversimplification of loss metrics in TransE, which limits its ability to effectively model complex entities and relations. TranSparse [25] introduces sparsity to construct an adaptive sparse matrix, addressing heterogeneity and imbalance issues in relation prediction. TransNS [13] enhances semantic affinity by selecting relevant neighbors as entity attributes and selecting negative triplets during the learning process to strengthen semantic interactions.

All these methods are based on deterministic vector spaces, overlooking the uncertainty between entities and relations. KG2E [22] adopts a density-based embedding approach, explicitly modeling entity and relation determinism in multi-dimensional Gaussian distribution space. Unlike TransE's symmetric metric approach, KG2E uses KL divergence to score triplets, effectively modeling various relation types. However, KG2E does not exhibit significant advantages in handling $(n - n)$ relations, as it does not consider entity types. Additionally, TransG [56] uses mixed Gaussian distributions to represent the diversity and uncertainty of relations, offering fine-grained categorization of relations, with significant practical application value. Other methods such as RotatE [45] and TorusE [14] share similarities with the TransE concept but employ different mapping operations. RotatE uses rotational operations, while TorusE employs torus embedding space to address regularization issues.

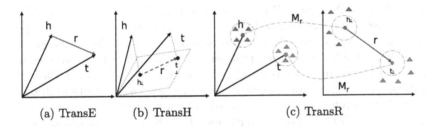

(a) TransE   (b) TransH   (c) TransR

Fig. 1. Simple illustrations of TransE, H, R model

II. Semantic Matching-Based. Semantic matching model employ similarity-based scoring functions to measure the plausibility of facts by comparing the latent semantics of entities and relations in their vector representations. This field has witnessed the emergence of several models that have evolved continuously in handling knowledge graph relation prediction tasks. One of the pioneering models is RESCAL [36], which adopts a tensor factorization-based approach. Each relation was associated with a relation matrix M_r to describe the semantic connections between entities and relations. However, RESCAL is afflicted by the computational overhead stemming from the necessity of maintaining individual relation matrices for each relation. To address the computational cost issue, DisMult [57] was introduced later, constraining the relation matrix to be a diagonal matrix, thereby reducing the number of model relation parameters. Nevertheless, it was limited to handling symmetric relations. Subsequently, HolE [35] combined the simplicity of DisMult with the expressiveness of RESCAL by introducing circular correlation as a composition operator. This allowed for a better capture of rich interactions between entities and relations, enhancing the model's representational power. ComplEx [47] further extended this idea by introducing complex embeddings to handle various semantic associations between different head and tail entities, including asymmetric relations. But the use of complex numbers added computational complexity challenges. In addition to tensor factorization-based methods, there are neural network-based models like SME [4], MLP [12], NTN [43], and NAM [30]. These models map entities and relations into vector embeddings, providing greater flexibility in semantic matching.

3.3 Deep Learning-Based Methods

Deep learning-driven relation prediction tasks rely on deep neural network or graph neural network. They work by learning embedded representations of entities and relations to predict unknown relation triplets. The strength of these methods lies in their ability to effectively capture intricate associations between entities and relations, suited for handling large-scale and high-dimensional relation data.

I. Deep Neural Network-Based. As convolutional neural networks (CNNs) have achieved significant success in fields such as image recognition [23], speech processing [62], and natural language processing [60], there has been an exploration of using CNNs for learning representations of entities and relations in the context of relation prediction tasks. ConvE [11] primarily utilizes 2D convolution operations to learn intricate associations between entities and relations. It offers a computational advantage due to its relatively lower parameter count compared to DisMult and R-GCN [41]. Nevertheless, ConvE focuses on capturing local relations across different dimensions among entities, lacking a comprehensive understanding of global relations. On the other hand, ConvKB [34] employs 1D convolution to maintain the translation properties of the TransE model, enabling better capturing of global relations and uniqueness between entities. Additionally, ConvKB simplifies the model structure by avoiding complex operations like matrix reshaping.

II. Graph Neural Network-Based. Graph neural network model, primarily based on graph convolution, leverage the structural information of graphs to learn representations of entities and relations. In this framework, entities and relations are treated as nodes and edges in a graph. These models predict unknown relations by passing information and performing graph convolution operations within the graph.

Graph Convolutional Network (GCN), which employs convolution operations to propagate node information. However, GCN has limitations when dealing with graphs with multiple relation. Relational Graph Convolutional Network (R-GCN) can be viewed as an autoencoder consisting of an encoder and a decoder. The encoder generates latent feature representations of entities, and the decoder scores triplets based on the representations generated by the encoder. R-GCN improves upon GCN by introducing relation-specific weights to handle graphs with multiple relations, with each relation having a different weight matrix. Weighted Graph Convolutional Network (WGCN) [42], building on R-GCN, introduces learnable attention weights to represent relations. It also adjusts the scope of required neighborhood information during local GCN aggregation, enhancing the embedding representations of graph nodes and relations. Vectorized Relational Graph Convolutional Network (VR-GCN) [59] addresses multi-relational networks by representing relations as relation vector embeddings. It introduces variational autoencoders (VAE) on top of R-GCN for probabilistic modeling, better handling incomplete knowledge graph data and uncertainty. Composite Graph Convolutional Network (CompGCN) [48] uses combination operators to jointly embed entities and relations in a relational graph, incorporating relations into consideration and updating them using matrix transformations.

3.4   Large Language Model-Based Methods

Relation prediction tasks based on large language model involve guiding the model's predictions through the design of prompts. These methods leverage the model's comprehension abilities to parse natural language and generate corresponding relations, rendering them particularly valuable in low-data resource scenarios. Approaches like LAMA [38] and LPAQA [26] treat LLMs as flexible knowledge bases. They utilize flexible prompts, such as ("Obama was born in,") to query information, yielding responses like ("Hawaii."). However, these queries generate results for tail entities and do not extend to a complete knowledge graph.

COMET [6], a commonsense relation prediction model fine-tuned on GPT-2 [15], has a limitation–it can only extract relations explicitly mentioned in the text. Symbolic Knowledge Distillation (SKD) [54] directly extracts knowledge from GPT-3 [16], but it heavily relies on GPT-3, and its quality discriminator is trained on existing knowledge graph data. This restricts its applicability to unknown relations. BertNet [20] introduces the concept of incorporating relations definitions into the input, such as prompts and example entity pairs. It also incorporates an efficient search and re-ranking mechanism that automatically retrieves and extracts high-quality knowledge related to the desired relations within LLMs. This approach is applicable to previously unseen relations.

Large language models empower relation prediction by extending their predictive capabilities to unknown relations, making them particularly valuable in low-data resource scenarios. In the future, researchers can explore more zero-shot or few-shot relation prediction solutions based on large language models and generalize them to address unknown relations. The description provided above illustrates the progressive relations among some of these models, Fig. 2 offering a more intuitive understanding of the connections between them.

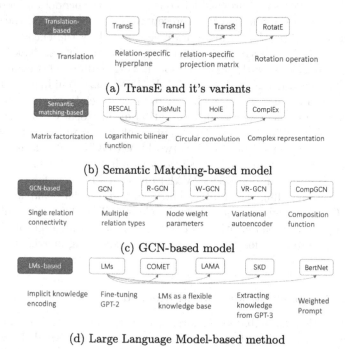

(a) TransE and it's variants

(b) Semantic Matching-based model

(c) GCN-based model

(d) Large Language Model-based method

Fig. 2. The evolution relations of models in different categories

3.5   Transductive and Inductive-Based Methods

Most models are in a transductive setting, which emphasizes the modeling and updating of entity information, often overlooking the crucial aspects of modeling and updating relations. Therefore, researchers have redirected their focus towards inductive models.

Inductive setting necessitates model to give heightened attention to relation information that remains independent of specific entities. Moreover, it requires them to grasp the underlying logical rules that govern the knowledge graph. GraphSAGE [19] introduced embedding vector representations tailored for inductive learning, but it relies on the presence of node features, which may not be available in many knowledge graphs. GraIL [46] takes a unique approach by deducing local substructures through the extraction of surrounding subgraphs encompassing target triples, showcasing a robust inductive bias. This approach enables the learning of relation semantics independently of entities and even facilitates the capture of valuable first-order logical rules through subgraph analysis. TACT [7] extends the capabilities of GraIL by introducing topological pattern modes for subgraphs, which enhances the learning of semantic associations between relations. CoMPILE [31], on the other hand, bolsters the message-passing process of GraIL by employing a communication kernel function, thereby intensifying the interaction between entities and relations.

Researchers have proposed hybrid approaches to fully leverage the advantages of rule-based and GCN-based methods. ExpressGNN [63] takes a more explicit route by incorporating prior logical rules directly into Graph Neural Networks (GNNs). This approach merges Markov Logic Networks (MLN) and GNN within the framework of variational EM [33]. pGAT [21] introduces a probabilistic logical graph attention network that performs inference by combining first-order logic and graph attention networks. Lastly, ConGLR [28] effectively models relations and performs logical reasoning by considering context graphs for subgraphs. Table 1 provides a summary of both transductive and inductive models.

Table 1. Summary of representative transductive and inductive models

Classification	Subtype	Models
Rule learning	Inductive	NELL [37], MLN [39], AMIE [18], AMIE+ [17], TensorLog [10], Neural LP [58], DRUM [40], RuleN [32] LAN [51]
Representation Learning	Transdutive	TransE [5], TransH [53], TransR [29], RESCAL [36], DisMult [57], ComplEx [47]
	Inductive	LAN [51], GraphSAGE [19]
Deep Learning	Transdutive	SME [4], NTN [43], MLP [12] ConvE [11], ConvKB [34], R-GCN [41] WGCN [42], VR-GCN [59], CompGCN [48]
	Inductive	GraIL [46], TACT [7], ExpressGNN [63], CoMPILE [31], pGAT [21], ConGLR [28]
Large Language Model	Inductive	LAMA [38], LPAQA [26], COMET [6], SKD [54], GPT-3 [16], BertNet [20]

4   Applications

Relation prediction is widely applied in various domains, including but not limited to recommendation systems, natural language understanding, drug discovery [61], financial forecasting [8], and medical diagnosis [27]. In the field of recommendation systems, relation prediction is used to establish connections between users and products or services, thereby enhancing the accuracy of recommendations. In the field of natural language processing, it aids machines in understanding semantic relations within text, benefiting applications like text analysis, information retrieval, and automated question-answering systems. Bioinformatics relies on relation prediction to forecast protein interactions, deepening our understanding of biological processes. In finance, it's pivotal for predicting stock market trends, currency exchange rate fluctuations, and providing vital decision support for investors. In medical diagnosis, it assists in disease diagnosis, predicts disease progression, and tailors personalized treatment plans.

With the emergence of large language models like GPT-3, the field of relation prediction applications has been significantly propelled. These models, with their strong natural language understanding and generation abilities, have become a driving force in advancing relation prediction tasks. They excel in extracting relation information from extensive text data when guided by well-designed prompts, thus facilitating emerging tasks like zero-shot and few-shot learning in relation prediction. In RelationPrompt [9], researchers can design prompts tailored to specific relation extraction tasks, making it possible to extract relation from vast amounts of text data efficiently. Furthermore, LLMs capability to convert unstructured natural language text into organized relation data significantly boosts the effectiveness and precision of relation prediction tasks. This

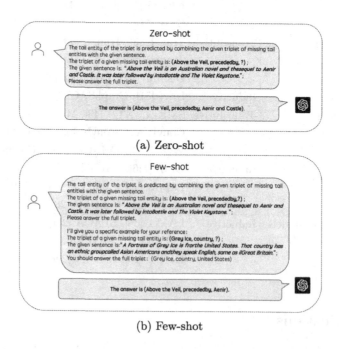

(a) Zero-shot

(b) Few-shot

Fig. 3. Examples of Relation Prediction with LLMs

synergy between large language models and relation prediction has opened up new possibilities and improved the overall performance of these applications.

In the example of Fig. 3: (Above the Veil, preceded by, ?), the target tail entity is Aneir. In the zero-shot setting, LLMs fails to comprehend the relation properly, leading to an incorrect response of Aenir and Castle. However, when the demonstration is incorporated, LLMs can successfully identify the target tail entity.

5 Conclusion

This comprehensive review examines various approaches to knowledge graph relation prediction, including rule-based, representation learning, deep learning, graph neural networks, and methods based on large language models, classifying them into transductive and inductive categories. Each approach has its strengths and limitations. Rule-based methods offer interpretability and logical rigor but face challenges with large and complex knowledge graphs. Representation learning methods provide vector embeddings for capturing complex patterns but are limited to transductive settings and struggle with unknown entities. Deep learning and graph neural network methods show promise in capturing complex graph patterns, while large language models excel in natural language understanding and reasoning. Inductive relation prediction is crucial for adapting to evolving knowledge graphs. Future research should explore flexible rule learning methods, combining rule-based reasoning with graph neural networks, and leverage large language models for enhanced knowledge understanding.

Acknowledgement. This work was supported by National Key R&D Program of China(2022YFB3103700) and Guangzhou Research Foundation (No. 202201020131).

References

1. Arora, S.: A survey on graph neural networks for knowledge graph completion. arXiv preprint arXiv:2007.12374 (2020)
2. Auer, S., Bizer, C., Kobilarov, G., Lehmann, J., Cyganiak, R., Ives, Z.: DBpedia: a nucleus for a web of open data. In: Aberer, K., et al. (eds.) ASWC/ISWC -2007. LNCS, vol. 4825, pp. 722–735. Springer, Heidelberg (2007). https://doi.org/10.1007/978-3-540-76298-0_52
3. Bollacker, K., Evans, C., Paritosh, P., Sturge, T., Taylor, J.: Freebase: a collaboratively created graph database for structuring human knowledge. In: SIGMOD, pp. 1247–1250 (2008)
4. Bordes, A., Glorot, X., Weston, J., Bengio, Y.: A semantic matching energy function for learning with multi-relational data: application to word-sense disambiguation. Mach. Learn. **94**, 233–259 (2014)
5. Bordes, A., Usunier, N., Garcia-Duran, A., Weston, J., Yakhnenko, O.: Translating embeddings for modeling multi-relational data. In: NIPS, vol. 26 (2013)
6. Bosselut, A., Rashkin, H., Sap, M., Malaviya, S., Celikyilmaz, A., Choi, Y.: COMET: commonsense transformers for knowledge graph construction. In: ACL, pp. 2–9 (2019)

7. Chen, J., He, H., Wu, F., Wang, J.: Topology-aware correlations between relations for inductive link prediction in knowledge graphs. In: AAAI, vol. 35, pp. 6271–6278 (2021)
8. Cheng, D., Yang, F., Wang, X., Zhang, Y., Zhang, L.: Knowledge graph-based event embedding framework for financial quantitative investments. In: SIGIR, pp. 2221–2230 (2020)
9. Chia, Y.K., Bing, L., Poria, S., Si, L.: RelationPrompt: leveraging prompts to generate synthetic data for zero-shot relation triplet extraction. In: ACL, pp. 45–57 (2022)
10. Cohen, W.W.: TensorLog: a differentiable deductive database. arXiv preprint arXiv:1605.06523 (2016)
11. Dettmers, T., Minervini, P., Stenetorp, P., Riedel, S.: Convolutional 2D knowledge graph embeddings. In: AAAI, vol. 32 (2018)
12. Dong, X., et al.: Knowledge vault: a web-scale approach to probabilistic knowledge fusion. In: SIGKDD, pp. 601–610 (2014)
13. Du, Z., Du, Z., Wang, L.: Open knowledge graph representation learning based on neighbors and semantic affinity. CSCD **52**(12), 2549–2561 (2019)
14. Ebisu, T., Ichise, R.: TorusE: knowledge graph embedding on a lie group. In: AAAI, vol. 32 (2018)
15. Ethayarajh, K.: How contextual are contextualized word representations? Comparing the geometry of BERT, ELMO, and GPT-2 embeddings. arXiv preprint arXiv:1909.00512 (2019)
16. Floridi, L., Chiriatti, M.: GPT-3: its nature, scope, limits, and consequences. Minds Mach **30**, 681–694 (2020)
17. Galárraga, L., Teflioudi, C., Hose, K., Suchanek, F.M.: Fast rule mining in ontological knowledge bases with AMIE+. VLDBJ **24**(6), 707–730 (2015)
18. Galárraga, L.A., Teflioudi, C., Hose, K., Suchanek, F.: AMIE: association rule mining under incomplete evidence in ontological knowledge bases. In: WWW, pp. 413–422 (2013)
19. Hamilton, W., Ying, Z., Leskovec, J.: Inductive representation learning on large graphs. In: NIPS, vol. 30 (2017)
20. Hao, S., et al.: BERTNet: harvesting knowledge graphs with arbitrary relations from pretrained language models. In: ACL, pp. 5000–5015 (2023)
21. Harsha Vardhan, L.V., Jia, G., Kok, S.: Probabilistic logic graph attention networks for reasoning. In: WWW, pp. 669–673 (2020)
22. He, S., Liu, K., Ji, G., Zhao, J.: Learning to represent knowledge graphs with Gaussian embedding. In: CIKM, pp. 623–632 (2015)
23. Hijazi, S., Kumar, R., Rowen, C., et al.: Using convolutional neural networks for image recognition. CDNS **9**(1) (2015)
24. Ji, G., He, S., Xu, L., Liu, K., Zhao, J.: Knowledge graph embedding via dynamic mapping matrix. In: ACL, pp. 687–696 (2015)
25. Ji, G., Liu, K., He, S., Zhao, J.: Knowledge graph completion with adaptive sparse transfer matrix. In: AAAI, vol. 30 (2016)
26. Jiang, Z., Xu, F.F., Araki, J., Neubig, G.: How can we know what language models know? TACL **8**, 423–438 (2020)
27. Li, L., et al.: Real-world data medical knowledge graph: construction and applications. AIDS Patient Care STDS **103**, 101817 (2020)
28. Lin, Q., et al.: Incorporating context graph with logical reasoning for inductive relation prediction. In: SIGIR, pp. 893–903 (2022)
29. Lin, Y., Liu, Z., Sun, M., Liu, Y., Zhu, X.: Learning entity and relation embeddings for knowledge graph completion. In: AAAI, vol. 29 (2015)

30. Liu, Q., et al.: Probabilistic reasoning via deep learning: neural association models. arXiv preprint arXiv:1603.07704 (2016)
31. Mai, S., Zheng, S., Yang, Y., Hu, H.: Communicative message passing for inductive relation reasoning. In: AAAI, vol. 35, pp. 4294–4302 (2021)
32. Meilicke, C., Fink, M., Wang, Y., Ruffinelli, D., Gemulla, R., Stuckenschmidt, H.: Fine-grained evaluation of rule- and embedding-based systems for knowledge graph completion. In: Vrandečić, D., et al. (eds.) ISWC 2018. LNCS, vol. 11136, pp. 3–20. Springer, Cham (2018). https://doi.org/10.1007/978-3-030-00671-6_1
33. Neal, R.M., Hinton, G.E.: A view of the EM algorithm that justifies incremental, sparse, and other variants. In: Jordan, M.I. (eds.) Learning in Graphical Models, vol. 89, pp. 355–368. Springer, Dordrecht (1998). https://doi.org/10.1007/978-94-011-5014-9_12
34. Nguyen, D.Q., Nguyen, T.D., Nguyen, D.Q., Phung, D.: A novel embedding model for knowledge base completion based on convolutional neural network. arXiv preprint arXiv:1712.02121 (2017)
35. Nickel, M., Rosasco, L., Poggio, T.: Holographic embeddings of knowledge graphs. In: AAAI, vol. 30 (2016)
36. Nickel, M., Tresp, V., Kriegel, H.P., et al.: A three-way model for collective learning on multi-relational data. In: ICML, vol. 11, pp. 3104482–3104584 (2011)
37. Paulheim, H., Bizer, C.: Improving the quality of linked data using statistical distributions. IJSWIS **10**(2), 63–86 (2014)
38. Petroni, F., et al.: Language models as knowledge bases? arXiv preprint arXiv:1909.01066 (2019)
39. Richardson, M., Domingos, P.: Markov logic networks. Mach. Learn. **62**, 107–136 (2006)
40. Sadeghian, A., Armandpour, M., Ding, P., Wang, D.Z.: DRUM: end-to-end differentiable rule mining on knowledge graphs. In: NIPS, vol. 32 (2019)
41. Schlichtkrull, M., Kipf, T.N., Bloem, P., van den Berg, R., Titov, I., Welling, M.: Modeling relational data with graph convolutional networks. In: Gangemi, A., et al. (eds.) ESWC 2018. LNCS, vol. 10843, pp. 593–607. Springer, Cham (2018). https://doi.org/10.1007/978-3-319-93417-4_38
42. Shang, C., Tang, Y., Huang, J., Bi, J., He, X., Zhou, B.: End-to-end structure-aware convolutional networks for knowledge base completion. In: AAAI, vol. 33, pp. 3060–3067 (2019)
43. Socher, R., Chen, D., Manning, C.D., Ng, A.: Reasoning with neural tensor networks for knowledge base completion. In: NIPS, vol. 26 (2013)
44. Suchanek, F.M., Kasneci, G., Weikum, G.: YAGO: a core of semantic knowledge. In: WWW, pp. 697–706 (2007)
45. Sun, Z., Deng, Z.H., Nie, J.Y., Tang, J.: RotatE: knowledge graph embedding by relational rotation in complex space. arXiv preprint arXiv:1902.10197 (2019)
46. Teru, K., Denis, E., Hamilton, W.: Inductive relation prediction by subgraph reasoning. In: ICML, pp. 9448–9457. PMLR (2020)
47. Trouillon, T., Welbl, J., Riedel, S., Gaussier, É., Bouchard, G.: Complex embeddings for simple link prediction. In: ICML, pp. 2071–2080. PMLR (2016)
48. Vashishth, S., Sanyal, S., Nitin, V., Talukdar, P.: Composition-based multi-relational graph convolutional networks. arXiv preprint arXiv:1911.03082 (2019)
49. Vrandečić, D.: Wikidata: a new platform for collaborative data collection. In: WWW, pp. 1063–1064 (2012)
50. Wang, M., Qiu, L., Wang, X.: A survey on knowledge graph embeddings for link prediction. Symmetry **13**(3), 485 (2021)

51. Wang, P., Han, J., Li, C., Pan, R.: Logic attention based neighborhood aggregation for inductive knowledge graph embedding. In: AAAI, vol. 33, pp. 7152–7159 (2019)
52. Wang, Q., Mao, Z., Wang, B., Guo, L.: Knowledge graph embedding: a survey of approaches and applications. TKDE **29**(12), 2724–2743 (2017)
53. Wang, Z., Zhang, J., Feng, J., Chen, Z.: Knowledge graph embedding by translating on hyperplanes. In: AAAI, vol. 28 (2014)
54. West, P., et al.: Symbolic knowledge distillation: from general language models to commonsense models. arXiv preprint arXiv:2110.07178 (2021)
55. Xiao, H., Huang, M., Hao, Y., Zhu, X.: TransA: an adaptive approach for knowledge graph embedding. arXiv preprint arXiv:1509.05490 (2015)
56. Xiao, H., Huang, M., Hao, Y., Zhu, X.: TransG: a generative mixture model for knowledge graph embedding. arXiv preprint arXiv:1509.05488 (2015)
57. Yang, B., Yih, W.T., He, X., Gao, J., Deng, L.: Embedding entities and relations for learning and inference in knowledge bases. arXiv preprint arXiv:1412.6575 (2014)
58. Yang, F., Yang, Z., Cohen, W.W.: Differentiable learning of logical rules for knowledge base reasoning. In: NIPS, vol. 30 (2017)
59. Ye, R., Li, X., Fang, Y., Zang, H., Wang, M.: A vectorized relational graph convolutional network for multi-relational network alignment. In: IJCAI, pp. 4135–4141 (2019)
60. Yin, W., Kann, K., Yu, M., Schütze, H.: Comparative study of CNN and RNN for natural language processing. arXiv preprint arXiv:1702.01923 (2017)
61. Zeng, X., Tu, X., Liu, Y., Fu, X., Su, Y.: Toward better drug discovery with knowledge graph. Curr. Opin. Struct. Biol. **72**, 114–126 (2022)
62. Zhang, Y., Chan, W., Jaitly, N.: Very deep convolutional networks for end-to-end speech recognition. In: ICASSP, pp. 4845–4849. IEEE (2017)
63. Zhang, Y., et al.: Efficient probabilistic logic reasoning with graph neural networks. arXiv preprint arXiv:2001.11850 (2020)
64. Zhang, Y., Qian, S., Fang, Q., Xu, C.: Multi-modal knowledge-aware attention network for question answering. J. Comput. Res. Dev. **57**(5), 1037–1045 (2020)

Negation: An Effective Method to Generate Hard Negatives

Yaqing Sheng, Weixin Zeng, and Jiuyang Tang[✉]

Laboratory for Big Data and Decision, National University of Defense Technology,
Changsha 410073, China
{shengyaqing,jiuyang_tang}@nudt.edu.cn

Abstract. Reasoning commonsense knowledge is essential for Artificial Intelligence, which requires high-quality commonsense knowledge. Recently, much progress has been made in automatic commonsense knowledge generation. However, most of the works focus on obtaining positive knowledge and lack negative information. Only a few works capture the importance of negative statements, but they struggle to produce high-quality knowledge. Although some efforts have been made to generate negative statements, they fail to consider the taxonomic hierarchy between entities and are not generally applicable, leading to the generation of low-quality negative samples. To resolve the issue, we put forward Negation, a framework for effectively generating hard negative knowledge. For each entity in the commonsense knowledge base, congeners are identified with hierarchical and semantic information. Then, negative candidates are produced by replacing the entity with congeners in each triple. In order to make negative knowledge more confusing and avoid false positive examples, we design two filtering steps to remove the amount of meaningless candidates. We empirically evaluate our proposed method Negation on the downstream task, and the results demonstrate that Negation and its components effectively help generate high-quality negative knowledge.

Keywords: Commonsense Knowledge · Commonsense Knowledge Base · Information Extraction

1 Introduction

Commonsense knowledge is the key to constructing powerful Artificial Intelligence (AI). There are several commonsense knowledge bases (CSKBs), such as ConceptNet [1], ATOMIC [2], ASCENT [3], ASCENT++ [4]. Recent studies focus on the acquisition of positive knowledge [4–6], such as extracting the positive triple (lion, ISA, carnivore) from the text "lions are carnivores" [4] and the completion of unseen facts based on existing knowledge graphs (KGs) [5,6]. However, knowing "not to do" is also important for humans and models. Humans avoid mistakes through negative commonsense. Negative statements can also help machine learning models to learn better, thereby improving the performance of the model.

However, only a few works capture the importance of negative statements. Safavi et al. [7] first provide the rigorous definition of negative knowledge in CSKB. They generate negative knowledge in widely used ConceptNet [1]. Given a CSKB and a pre-trained language model (LM), the LM is first fine-tuned with CSKB. Then, negative candidates are generated by replacing the head/tail entity of a triple with dense k-nearest neighbors retrieval. In the final step, negative candidates are ranked by the fine-tuned LM [7]. However, the relation in ConceptNet [1] is pre-defined. For more abundant semantics in ASCENT++ [4], Arnaout et al. [8] propose a series of natural language steps to generate negative knowledge. The comparable concepts are identified with pre-computing embeddings and taxonomic checks. Positive statements of comparable concepts become negative candidates. The candidates are then scrutinized and ranked by informativeness. Considering the powerful generation ability of Large language models (LLMs), negative knowledge is also generated from LLMs [9,10].

Although these methods are able to produce an amount of negative knowledge, they suffer some limitations. First, the taxonomic hierarchy between entities has not been taken into account. Some methods do not consider the taxonomic hierarchy [7,9,10]. They do not or only use the pre-trained embeddings to produce similar entities, which merely seem similar in the embedding space. Arnaout et al. [8] consider this through an Is-A database. They filter the comparable concepts that do not share a homonym with the given concept. However, this database is large and the time cost of checking is high. Secondly, the quality of negative knowledge cannot be guaranteed. On the assumption of a local closed world [11], traditional methods [7,8] cannot effectively detect false positive examples that are out of CSKB. As for LLMs, which have excellent performance, generate ambiguous [9] and even false negative commonsense knowledge [10]. Finally, these methods are not generally applicable. NegatER [7] relies on the given CSKB having pre-defined relations, that cannot be identified in another CSKB [4,12]. For UnCommonSense [8], some steps cannot play a role in semi-structured triples, such as KB-based scoring and LM-based scoring. These steps are more inclined to explore the characteristics of natural language, and not fit for pre-defined phrases.

To deal with the problems mentioned above, we propose Negation, an effective method to generate high-quality negative commonsense knowledge. We exploit Box Embedding [13] to identify congeners. The box structure contains the implicit information of the hierarchy [13]. Also, Box Embedding includes the semantic information based on the pre-trained BERT [14]. Therefore, this method considers hierarchical and semantic information with low time cost calculation. Additionally, we apply double-check steps to obtain high-quality knowledge. Inspired by the recent work in negative knowledge generation [7], we use BERT [14] to obtain the most confusing negative triples. The given CSKB is fine-tuned to acquire the positive belief. Then the candidates are input to the fine-tuned LM. Through scoring and ranking the candidates with LM scores, the more confusing candidates are screened out. Meanwhile, we adopt powerful LLMs that perform well in polar yes-or-no questions [10] to scrutinize false positive candidates with "FALSE" answers. In addition, the whole process of Negation only employs entities in a given CSKB, with no relation used. Therefore, our proposed method Negation is not limited to the triple structure, because we do not consider whether the relation is pre-defined. We empirically Negation the downstream task in both embedding-based and graph-based baselines. The results demonstrate that Negation achieves superior performance and can produce high-quality negative knowledge.

Contribution The main contribution of this article can be summarized as:

(1) We adopt Box Embedding to obtain the hierarchical and semantic information between entities with low time cost.
(2) We design double-check steps, LM check and LLM check to make the quality of negative knowledge higher.
(3) We show the usefulness of our proposed model to produce high-quality negative knowledge via the downstream task.

2 Problem Definition

A CSKB consists of finite triples (h, r, t), where h is the head entity, t is the tail entity, and r refers to the relationship between h and r. Triples in the CSKB are positive or true knowledge, denoted as $^+(h, r, t)$ and the high-dimensional representation is $^+(X_h, X_r, X_t)$. As defined in NegatER [7], we denote a negative triple as $^-(h', r, t)$ and $^-(h, r, t')$. We follow the definition of the research problem in UnCommonSense [8]: Given a target concept s in a CSKB, generate a ranked list of truly negative and informative statements. Following the previous studies [7,8], we present higher requirements for the negative knowledge:

R1: Negative knowledge must be grammatically correct and semantically reasonable. Negative Knowledge is presented as triples, so negative knowledge must follow the grammatical rules under relation in the same way as positive knowledge. For example, the negative triple $^-(lion', AtLocation, forest)$ is correct, but the negative triple $^-(lion', HasA, forest)$ cannot satisfy the grammar rules under the relation "HasA".

R2: Avoid false positive negatives. Generally, KBs follow the open-world assumption. It means that out-of-KB knowledge is not always wrong. Negative knowledge generated by the given CSKB may be positive or true.

R3: Negative knowledge must be confusing. We expect that negative knowledge should give more information. Libai is a Poet Genius in China. For negative triples, $^-(Dufu', IsA, PoetGenius)$ and $^-(TimCook', IsA, PoetGenius)$, Libai and Dufu are the celebrated poets in the Tang Dynasty of China. However, Tim Cook is from America and a current person. Clearly, the first one provides more information than the second one. Therefore, the second one is more confusing. This highly confusing negative knowledge is called high-quality negative knowledge.

3   Framework

We propose Negation, an automatic method to generate high-quality negative commonsense knowledge effectively. As shown in Fig. 1, Negation consists of three steps. At first, Negation retrieves the congeners of entities. Then, Negation generates initial negative candidates by replacing the head and tail entity of a triple with their congeners. Furthermore, in order to obey the grammatical rules, Negation discards candidates whose entity-relation pairs do not appear in the given CSKB. To make negative knowledge more confusing, we use a fine-tuned LM to filter and sort negative candidates. Taking the effectiveness of LLMs in answering polar yes-or-no questions [10], we scrutinize false positive candidates with question-answering tasks.

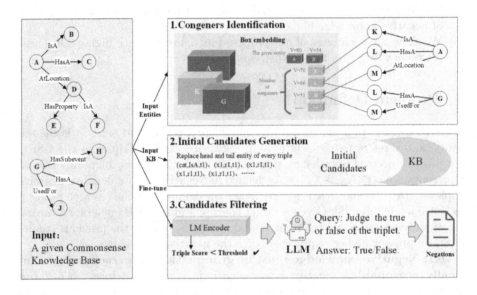

Fig. 1. Architecture of Negation

3.1   Congeners Identification

Out-of-KB negatives are extremely large and most of them are meaningless. Intuitively, a local closed-world assumption [11] is made when generating congeners. It is considered that congeners obtained from a given KB would have a stronger correlation with existing entities. For example, for the given entity "lion", the negative statements are likely to come from entities that are animals, such as tigers, leopards, and whales, rather than tables and chairs.

To generate congeners, there are several methods, such as graph-based methods [15], semantic-based methods [14,16]. Congeners generated based on graph structure ignore the semantic information of entities. However, pre-trained embeddings do not consider the hierarchy of the entity. For instance, "trunk" is highly related to "elephant" in semantic embeddings, but they are not similar actually [8]. If the information 'trunk' and 'elephant' are not the same category" is applied, such mistakes would be avoided. Therefore, to balance hierarchical and semantic information, we use Box Embedding [17] to detect congeners. Box embeddings represent entities as n-dimensional hyperrectangles [13]. A box x is characterized by two points (x_m, x_M), where $x_m, x_M \in R^d$ are the minimum and the maximum corners of the box x, $x_{m,i} \leq x_{M,i}$, i for each coordinate $i \in \{1, ..., d\}$.

The box embedding calculates the minimum and the maximum corners of the box by pre-trained BERT [14] to obtain the representation of each box. The box structure contains the implicit information of the hierarchy [13]. At the same time, box embeddings include the semantic information based on the pre-trained BERT [14]. Therefore, box embeddings include two types of information, hierarchical and semantic information. The most intuitive representation of a box is volume. We use box volume as the basis for calculating congeners. The volume of the box x, $Vol(x)$ is computed as

$$Vol(x) = \prod_i (x_{m,i} - x_{M,i}) \tag{1}$$

Through calculating [18], we obtain the box volume with two types of information. Intuitively, we assume that the box volume difference represents the similarity of two entities. As shown in Fig. 1, the box volume of each entity is acquired by calculating between the given entities X are candidates as candidates. The congeners with the smallest volume difference are selected as candidates.: $\{X : [X_1, ..., X_k] | X_i \in min(X - X_i)\}$.

3.2   Initial Candidates Generation

To produce negative candidates, we replace the head and tail entity of a triple with their congeners that are identified in the first steps.

For a triple $^+(X_h, X_r, X_t)$, $\{X_h : [X_1, ..., X_k] | X_i \in min(X - X_i)\}$, replacing X_h, $\{X_t : [X_1, ..., X_k] | X_j \in min(X - X_j)\}$ replacing X_t, candidates could be obtained as follows.

$$Candidates1 = \{^-(X_i, X_r, X_t) \bigcup {}^-(X_h, X_r, X_j)\}, i, j = 1, ..., k \tag{2}$$

However, negative candidates generated by the above step might be inaccurate. To obey the grammatical rule A1, Negation discards candidates whose entity-relation pairs do not appear in the given CSKB.

$$Candidates2 = \{^-(X_i, X_r, X_t) \bigcup {}^-(X_h, X_r, X_j) \\ |(X_i, X_r), (X_r, X_J) \in KB\}, i, j = 1, ..., k \tag{3}$$

3.3   Candidates Filtering

Through the above steps, initial negative candidates have been produced. We meet R2 and R3 with LM check and LMM check.

LM Check. In order to make negative knowledge highly confusing, we follow the strategy that employs fine-tuned LM [19] to rank negative candidates [7]. This step fine tune the LM with the given CSKB. Then, acquire the negative candidates below the threshold θ. Most of the candidates are easy negatives, which is not enough to train a great model. The 5% of hard negatives really determine the performance of the model [20]. The lower score triples will be removed, and sorted by the LM score.

$$Candidates3 = \{LMscore(Candidates2) < \theta\} \qquad (4)$$

$$Candidates4 = \{Candidates3|RandPercent(Candidate) \geq 50\%\} \qquad (5)$$

LMM Check. According to Cai et al. [20], even a small percentage of false positives in the training examples can significantly affect the model performance. LLMs are trained on vast amounts of data and perform well various tasks, which scale is huge and contains hundreds of millions of parameters [21], such as GPT4 [22], Palm [23], Llama [24]. LLMs perform great in answering polar yes-or-no questions [10]. Therefore, we design a zero-shot prompt in LLM [25,26] to obtain the "TRUE" or "FALSE" answered by LLM.

$$Negatives = \{Candidates4|LLM(Candidate) == False\} \qquad (6)$$

4   Experiments

For negative knowledge generation, we employ the dataset produced by Li et al. [19], which consists of 100k/2400/2400 train/valid/test triple. The dataset contains 78334 entity phrases and 34 relations from ConceptNet [27].

4.1   Task Evaluation

Since there is no intuitive indicator to measure the pros and cons of negative knowledge, we apply it to downstream tasks. We adopt the triple classification task [7]. The efficiency of negative samples in this task is crucial.

To evaluate the effectiveness of negative knowledge, we use a more challenging dataset ConceptNet-TN, introduced by NegatER [7]. The dataset consists of 36210/3278/3278 train/validation/test triples, which come from ConceptNet [27]. The validation/test is retrieved from real negative knowledge (e.g.,

NOTISA, six total). We discovered that there is still a sample imbalance by analyzing the number of positive and negative samples. For this reason, we replace the binary cross-entropy loss in code [7] with focal loss [28].

The classes are balanced positive/negative, so accuracy is our main performance metric. Moreover, precision, recall, and F1 score are presented. We compare the following baseline for generating negative knowledge. We conduct multiple experiments for each method and calculate the average for the final results.

Baselines

- UNIFORM [19,29]: For each triple in the CSKB, we replace the head or tail entity by uniform sampling another entity from KB.
- COMET [30]: We add a "not" token to each head and then generate 10 entities using beam search. Finally, the tail entity is replaced with generated entities.
- ROTATE-SA [31]: For every triple, UNIFORM generated negative candidates. These candidates are ranked by RotatE KB embedding and take the top 50% as the final negatives.
- SANS [32]: We replace the head or tail entity of every triple with entities by sampling from the k-hop neighborhood. k is set to 2.
- NegatER-threshold [7]: This work proposes an unsupervised method that rank out-of-KB candidates with a fine-tuned LM.
- UnCommonSence [8]: This work designs a series of natural language steps to generate negative knowledge. We reproduce the code according to the paper. However, on account of the time cost and non-versatility, there are two steps not included, the taxonomical check and the LM-based scoring.

Analysis. As shown in Table. 1, the accuracy and F1 score of Negation are optimal, 0.83%-2.47% higher than the baselines. In the triple classification task, high-quality triples can improve the model performance. The results show that negatives generated by Negation significantly enhance the performance of the classifier. The most competitive model in the baselines is NegatER. The two indicators, accuracy and recall have the largest gap between NegatER and Negation. The formula of accuracy is $\frac{TP+TN}{TP+FP+TN+FN}$. The formula of recall is $\frac{TP}{TP+FN}$. From the formal, We argue that Negation can identify false positive examples better than NegatER.

The performance of ROTATE-SA and SANS are obviously lower than other baselines. These methods assume that the KB is a dense graph. In reality, CSKBs are mostly sparse. Additionally, K-hop neighborhood sampling provides limited information for nodes and cannot satisfy the semantic consistency with the given node. Consequently, the sampling method may generate plenty of false positive examples to damage the model.

UNIFORM and COMET achieved moderate results in the baseline. As for UNIFORM, its high recall and low accuracy indicate that positive examples can be predicted correctly. The model has stronger prediction ability for positive examples, but weak prediction ability for negative ones. We argue that most of the easy negatives are generated by UNIFORM, which cannot enhance the model performance. Negation ensures that the generated negatives are hard through Box Embedding and double checks.

Table 1. Results on ConceptNet-TN. **Bold**: Best result, <u>Underline</u>: Second-best result.

Method	Accuracy	Precision	Recall	F1
UNIFORM	76.32	73.19	**83.23**	<u>77.89</u>
COMET	76.27	**76.00**	76.89	76.44
ROTATE-SA	75.97	74.45	79.17	76.74
SANS	74.88	72.28	80.86	76.33
NegatER-threshold	76.52	<u>75.57</u>	78.44	76.98
UnCommonSense	75.83	74.25	79.34	76.17
Negation	**77.35**	75.49	<u>81.05</u>	**78.17**

4.2 Ablation Study

In the ablation study, Table. 2 reveals the influence of each component in generating high-quality negatives. In line 1 of Table. 2, we replace Box Embedding with a dense k nearest-neighbors retrieval approach. As shown in Table. 2, the accuracy drops the most after removing the Box Embedding step. A possible explanation for this might be that the quality of the replacement entity used to generate negative candidates greatly influences the outcome. Likewise, Box Embedding employed in Negation has the ability to maximize both hierarchical and semantic information, which is the basis of generating negative knowledge.

Table 2. Ablation study results. (-step) represents delete the step.

Method	Accuracy	Precision	Recall
Negation(-Box Embedding)	76.30	75.46	77.98
Negation(-LM check)	77.02	75.31	80.43
Negation(-LLM check)	77.19	75.91	79.73
Negation	77.35	75.49	81.05

5 Related Works

5.1 Commonsense Knowledge Bases

CSKB construction is a large-scale and long-term work. ConceptNet [1,27] is currently used and maintained CSKB, consisting of a head entity, tail entity, and pre-defined relation. ATOMIC [2] focuses on reasoning knowledge organized in if-then relationships. ASCENT++ [4] breaks the limitation of pre-defined relation, that extracts natural language triples from large-scale web text.

5.2   Negative Knowledge

About the negative knowledge in KB, ConceptNet [27] defines six negative relations and gives the corresponding negative triples. Quasimodo [12] includes 350k negative statements, but the quality of those is poor due to the extraction steps [8]. In recent years, there has been increased focus on the generation of negative knowledge. NegatER [7] presents fine-tuned LM to detect negative knowledge in the KB. UnCommonSense [8] proposes unique pipelines to retrieve the negatives in ASCENT++ [4]. In addition, Arnaout et al. [9] found that LLMs often generate ambiguous negative knowledge and lack the ability to generate high-quality negative knowledge through experimentation.

5.3   Large Language Models

ChatGPT is the most representative product in LLMs, and its emergence has brought the revolution of AI. LLMs are trained on a large amount of data and contain hundreds of millions of parameters [21]. LLMs will have the emergence capability when the model scale surpasses a certain value by continuously expanding the model scale [33]. The emergence capability makes a great improvement in LLMs, which enhances the performance of downstream tasks. Although LLMs have powerful generation abilities, they lack the capability to detect fake information in generated content [10, 34]. Thus, we only use the powerful question-and-answer ability of LLMs to eliminate negative candidates.

6   Conclusion

In the paper, we propose Negation, which considers hierarchical and semantic information to generate negative knowledge. Then high-quality negative knowledge is filtered by a series of steps. We demonstrate the effectiveness of Negation, which is optimal both in accuracy and F1 score. Possible research directions to explore involve generation negatives under different circumstances [35]. For instance, "dragon" is a symbol in Chinese traditional culture. However, in American films, "dragon" is an animal. Hence, the negatives in the two conditions are not the same. Furthermore, considering the fact that LLMs generate fake content [34], it is worth studying how to avoid false generation through existing negative knowledge.

References

1. Speer, R., Chin, J., Havasi, C.: Conceptnet 5.5: An open multilingual graph of general knowledge. In: Proceedings of the AAAI Conference on Artificial Intelligence. vol. 31 (2017)
2. Sap, M., et al.: Atomic: an atlas of machine commonsense for if-then reasoning. In: Proceedings of the AAAI Conference on Artificial Intelligence. vol. 33, pp. 3027–3035 (2019)

3. Nguyen, T.P., Razniewski, S., Weikum, G.: Advanced semantics for commonsense knowledge extraction. In: Proceedings of the Web Conference 2021, pp. 2636–2647 (2021)
4. Nguyen, T.P., Razniewski, S., Romero, J., Weikum, G.: Refined commonsense knowledge from large-scale web contents. IEEE Trans. Knowl. Data Eng. (2022)
5. Chen, Z., Wang, Y., Zhao, B., Cheng, J., Zhao, X., Duan, Z.: Knowledge graph completion: a review. IEEE Access 8, 192435–192456 (2020)
6. Lin, Q., Mao, R., Liu, J., Xu, F., Cambria, E.: Fusing topology contexts and logical rules in language models for knowledge graph completion. Inf. Fusion 90, 253–264 (2023)
7. Safavi, T., Zhu, J., Koutra, D.: Negater: unsupervised discovery of negatives in commonsense knowledge bases. arXiv preprint arXiv:2011.07497 (2020)
8. Arnaout, H., Razniewski, S., Weikum, G., Pan, J.Z.: Uncommonsense: Informative negative knowledge about everyday concepts. In: Proceedings of the 31st ACM International Conference on Information Knowledge Management, pp. 37–46 (2022)
9. Arnaout, H., Razniewski, S.: Can large language models generate salient negative statements? arXiv preprint arXiv:2305.16755 (2023)
10. Chen, J., Shi, W., Fu, Z., Cheng, S., Li, L., Xiao, Y.: Say what you mean! large language models speak too positively about negative commonsense knowledge. arXiv preprint. arXiv:2305.05976 (2023)
11. Galárraga, L., Razniewski, S., Amarilli, A., Suchanek, F.M.: Predicting completeness in knowledge bases. In: Proceedings of the tenth ACM International Conference on Web Search and Data Mining, pp. 375–383 (2017)
12. Romero, J., Razniewski, S., Pal, K., Z. Pan, J., Sakhadeo, A., Weikum, G.: Commonsense properties from query logs and question answering forums. In: Proceedings of the 28th ACM International Conference on Information and Knowledge Management, pp. 1411–1420 (2019)
13. Onoe, Y., Boratko, M., McCallum, A., Durrett, G.: Modeling fine-grained entity types with box embeddings. arXiv preprint. arXiv:2101.00345 (2021)
14. Devlin, J., Chang, M.W., Lee, K., Toutanova, K.: Bert: Pre-training of deep bidirectional transformers for language understanding. arXiv preprint. arXiv:1810.04805 (2018)
15. Ponza, M., Ferragina, P., Chakrabarti, S.: A two-stage framework for computing entity relatedness in wikipedia. In: Proceedings of the 2017 ACM on Conference on Information and Knowledge Management, pp. 1867–1876 (2017)
16. Pennington, J., Socher, R., Manning, C.D.: Glove: Global vectors for word representation. In: Proceedings of the 2014 Conference on Empirical Methods in Natural Language Processing (EMNLP), pp. 1532–1543 (2014)
17. Dasgupta, S., Boratko, M., Zhang, D., Vilnis, L., Li, X., McCallum, A.: Improving local identifiability in probabilistic box embeddings. Adv. Neural. Inf. Process. Syst. 33, 182–192 (2020)
18. Chheda, T., G.: Box embeddings: an open-source library for representation learning using geometric structures. arXiv preprint arXiv:2109.04997 (2021)
19. Li, X., Taheri, A., Tu, L., Gimpel, K.: Commonsense knowledge base completion. In: Proceedings of the 54th Annual Meeting of the Association for Computational Linguistics (Volume 1: Long Papers), pp. 1445–1455 (2016)
20. Cai, T.T., Frankle, J., Schwab, D.J., Morcos, A.S.: Are all negatives created equal in contrastive instance discrimination? arXiv preprint arXiv:2010.06682 (2020)
21. Zhao, W.X., et al.: A survey of large language models. arXiv preprint. arXiv:2303.18223 (2023)

22. OpenAI: Gpt-4 technical report (2023)
23. Anil, R., et al.: Palm 2 technical report. arXiv preprint. arXiv:2305.10403 (2023)
24. Touvron, H., et al.: Llama: open and efficient foundation language models. arXiv preprint arXiv:2302.13971 (2023)
25. Du, Z., et al.: GLM: General language model pretraining with autoregressive blank infilling. arXiv preprint. arXiv:2103.10360 (2021)
26. Zeng, A., et al.: GLM-130b: An open bilingual pre-trained model. arXiv preprint. arXiv:2210.02414 (2022)
27. Speer, R., et al.: Representing general relational knowledge in conceptnet 5. In: LREC. vol. 2012, pp. 3679–86 (2012)
28. Lin, T.Y., Goyal, P., Girshick, R., He, K., Dollár, P.: Focal loss for dense object detection. In: Proceedings of the IEEE International Conference on Computer Vision, pp. 2980–2988 (2017)
29. Saito, I., Nishida, K., Asano, H., Tomita, J.: Commonsense knowledge base completion and generation. In: Proceedings of the 22nd Conference on Computational Natural Language Learning, pp. 141–150 (2018)
30. Bosselut, A., Rashkin, H., Sap, M., Malaviya, C., Celikyilmaz, A., Choi, Y.: Comet: commonsense transformers for automatic knowledge graph construction. arXiv preprint. arXiv:1906.05317 (2019)
31. Sun, Z., Deng, Z.H., Nie, J.Y., Tang, J.: Rotate: Knowledge graph embedding by relational rotation in complex space. arXiv preprint arXiv:1902.10197 (2019)
32. Ahrabian, K., Feizi, A., Salehi, Y., Hamilton, W.L., Bose, A.J.: Structure aware negative sampling in knowledge graphs. arXiv preprint. arXiv:2009.11355 (2020)
33. Wei, J., et al.: Emergent abilities of large language models. arXiv preprint. arXiv:2206.07682 (2022)
34. Kaddour, J., Harris, J., Mozes, M., Bradley, H., Raileanu, R., McHardy, R.: Challenges and applications of large language models. arXiv preprint. arXiv:2307.10169 (2023)
35. Nguyen, T.P., Razniewski, S., Varde, A., Weikum, G.: Extracting cultural commonsense knowledge at scale. In: Proceedings of the ACM Web Conference 2023, pp. 1907–1917 (2023)

SemiBDMA 2023

Diversified Group Recommendation Model for Social Network

Dong Li[1], Zhenshuo Liu[1], Zhanghui Wang[1(✉)], Jin Liu[2], Yue Kou[2], and Lingling Zhang[1]

[1] School of Information, Liaoning University, Shenyang 110036, China
dongli@lnu.edu.cn
[2] School of Computer Science and Engineering, Northeastern University, Shenyang 110819, China

Abstract. Group recommendation can recommend satisfactory activities to group members in the recommendation system. In the research of group recommendation, the main issue is how to combine the preferences of different group members. Most of the existing group recommendations adopt a single aggregation strategy to aggregate the preferences of different group members, unable to fulfill the needs of diversified group decision-making. At the same time, most of these group recommendation methods rely on intuition or hypothesis to analyze the influence of group members, which lacks convincing theoretical support. To overcome this issue, we propose the Diversified Group Recommendation Model for social network (DGRM). This model considers two aspects of social choice and social influence, models the diversity of groups, and adopts different group recommendation strategies for different groups, which can better meet the diverse needs of users and groups. We propose a group recommendation strategy based on score fusion, which can better meet the diverse needs of users and groups. Firstly, a matrix factorization-based individual rating prediction method and a Bayesian model-based individual rating prediction method are proposed, respectively, to predict the individual ratings based on user-item interactions. Secondly, different strategies for scoring fusion are proposed, and group recommendations are made based on the fused scores for different types of groups. Finally, we verify the feasibility and effectiveness of the key technologies proposed in this paper by conducting experiments, which demonstrates the effectiveness of our proposed methods.

Keywords: Group Recommendation · Social Choice · Social Influence · Matrix Factorization · Scoring Fusion

1 Introduction

With the rapid development of the Internet information age, there is now an increasing demand for people to find suitable target items. In order to solve this problem, people put forward the concept of a recommendation system [1]. The function of a recommendation system is to make relevant predictions about the relationship between users

© The Author(s), under exclusive license to Springer Nature Singapore Pte Ltd. 2024
X. Song et al. (Eds.): APWeb-WAIM Workshops 2023, CCIS 2094, pp. 39–51, 2024.
https://doi.org/10.1007/978-981-97-2991-3_4

and items, as well as users' "preferences" for items. When users rate items, scoring is a common method used to provide feedback. Users select appropriate scores from the rating indicators provided by the system [2].

In recent years, the target users of recommendation systems have largely been individuals [3]. However, in many scenarios, there is a growing need to provide recommendation information to specific groups. By obtaining information about group users and understanding the characteristics of various scenarios, group recommendation systems can offer clear and rapid recommendation schemes tailored for these group users.

This paper proposes the Diversified Group Recommendation Model for social network (DGRM). This model considers two aspects of social choice and social influence, models the diversity of groups, and adopts different group recommendation strategies for different groups, which can better meet the diverse needs of users and groups. More specifically, we make the following contributions.

(1) We propose a novel group recommendation strategy based on score fusion, which can better meet the diverse needs of users and groups. A matrix factorization-based individual rating prediction method and a Bayesian model-based individual rating prediction method are proposed, respectively, to predict the individual ratings based on user-item interactions.
(2) The different strategies for scoring fusion are proposed, and group recommendations are made based on the fused scores for different types of groups.
(3) We verify the feasibility and effectiveness of the key technologies proposed in this paper by conducting experiments, which demonstrates the effectiveness of our proposed methods.

2 Related Works

The emphasis of domestic recommendation systems is on personalized recommendation, and correspondingly, group recommendation systems are very underdeveloped. Moreover, most recommendation systems serve specific scenarios, such as PolyLens [4], Travel Decision Forum [3], INTRIGUE [5], TV4MW [6], and others [7].

The most difference between group recommendation and individual recommendation is that the final decision may not be made by an individual, but rather the result of negotiations among group members [7]. For group recommendation, the key challenge is that the general recommendation system generates a personalized recommendation list, and how to integrate this list into a group recommendation effectively. For example, the fusion method adopted by the MusicFX system is an average strategy that causes no pain [8].

Researchers have conducted many large-scale explorations in real social networks to examine the mechanisms of social choice and social influence within networks. For instance, some survey results indicate that social choice plays a more prominent role on Flickr and Facebook [9], while the social influence mechanism is stronger on Youtube, Epinions, and ScienceNet. Moreover, both social choice and social influence mechanisms play a role in some paper bookmarking platforms.

The two most prevalent recommendation system types are collaborative filtering (CF) and content-based recommendation. Collaborative filtering relies on the preferences of a subset of users within a group and the degree of preference for specific items. That is, it predicts the preferences of target users by utilizing the objects associated with them, which can be considered as "the wisdom of the group" [10]. In contrast, content-based recommendation systems primarily focus on the item and suggest relevant goals based on the characteristics of the item [11].

Compared to the collaborative filtering recommendation model, a memory-based recommendation system may not always be as fast or scalable, especially in the context of a real-time system generating recommendations based on very large datasets. To achieve these goals, a model-based recommendation system [12] is employed.

3 Model Overview

In this section, we propose the Diversified Group Recommendation Model for social network (DGRM), which considers two aspects of social choice and social influence. The method divides users into positive users and conformity users, and further categorizes groups into positive group and conformity group. The diversified modeling is applied to each of these categories.

3.1 DGRM Model

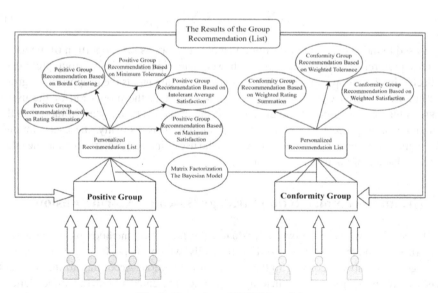

Fig. 1. Overview of DGRM model

Figure 1 illustrates the framework of our DGRM model. The model consists of two types of user groups which are positive group and conformity group, respectively. The positive group consists of five users, indicating that it has a higher number of positive users. And the conformity group consists of three users, indicating that it has a larger number of positive users.

Individual score prediction based on matrix factorization and Bayesian personalized ranking are performed for users, respectively, resulting in personalized recommendation lists for each user. These lists are subsequently used to merge groups of different strategies.

When devising a group recommendation strategy, various recommendation methods are used to employed depending on the type of group. For the positive group, five types of group recommendation strategies are available, which are those based on rating summation, borda counting, minimum tolerance, intolerance average satisfaction, and maximum satisfaction.

For the conformity group, three recommendation strategies are available, which are weighted rating summation, weighted tolerance, and weighted satisfaction.

3.2 Diversified Clustering Modeling

In this paper, the interest preference of the user i is expressed as p_i, the group preference of the group g to which user i joins is expressed as p_g. User interest preference is that users may join multiple groups due to various factors, leading to differing external performance in different groups. As a result, the user's interest preference can be calculated (shown as (1)).

$$p_i = \sum w_{ic} \cdot p_{ic} + \sum w_{gc} \cdot p_g \tag{1}$$

p_i is used to indicate the interest preference of user i, w_{ic} is the proportion of the user's interest preference in the group c in which user i participates, p_{ic} is the group preference reflected by the group c in which user i participates, and w_{gc} is the proportion of group c's interest preference in the group. It can be seen that the proportion of the user's own preference and group preference is 1. If the group exerts a higher influence in the communication process, the greater the proportion of w_{gc}. Finally, the user's interest preference p_i depends on two aspects. One is the user's interest preference, and the other is the influence of the user in the group.

4 Group Recommendation Strategy Based on Scoring Fusion

In this section, we propose two individual score prediction methods based on matrix factorization and the bayesian model, respectively, which can predict the item scores of each group member according to the user-item interaction. According to the different types of groups, various scoring fusion strategies are presented. Group recommendations are made based on the fused scores.

4.1 Individual Score Prediction Based on Matrix Factorization

Collaborative filtering is a widely used recommendation algorithm, among which the matrix factorization model is the most important one. Compared with other classical recommendation algorithms, the matrix factorization model has better performance. Matrix factorization model is a hidden semantic model [13], and its principle is to decompose an initial score prediction matrix R into two low-dimensional matrices with hidden factor dimensions and multiply them, user matrix M and item matrix N, which can be multiplied to obtain prediction scores, defined as (2).

$$r_{ij} \backsim p_i q_j^T = \sum p_{ik} \times q_{jk}^T \ll=\gg R \backsim M \times N^T \tag{2}$$

In the Eq. (2), K represents the number of eigenvalues implied in the user-item matrix. When it is used for group recommendation, the average satisfaction of users in the group is the highest, which is expressed as (3).

$$TOP(g, M) = argmax_i^M z_{gij} \tag{3}$$

M represents the top M best-recommended results of the group.

The individual preference defined in this paper can be expressed by Eq. (4) of matrix factorization.

$$z_{I,J,K} = W_i^I \sum I_{i,k} J_{i,k}^{IT} + W_g^C \sum g_{g,k} J_{j,k}^{GT} + \beth_I \tag{4}$$

In the Eq. (4), k is the number of implicit eigenvalues, $\beth_I$ which is the error degree of the item. The more popular the item, $\beth_I$ the greater the value may be.

This part adopts a personalized recommendation based on matrix factorization, and its process is shown in Algorithm 1.

Algorithm 1. Matrix Factorization Training Algorithm

Input: User operation behavior, score prediction matrix R
Output: The two low-dimensional prediction matrices P, Q
01. Initialize(R)   // Initialize the R matrix
02. For t =1, 2, 3 … Step do
03.   If $R[I][J] > 0$;   // Limit the score to be greater than 0
04. $P[i][k] = P[i][k] + alpha*(2*eij*Q[k][j]-beta*P[i][k])$ // Adding regex
05. $Q[k][j] = Q[k][j] + alpha*(2*eij*P[i][k]-beta*Q[k][j])$
06. $Loss$ = countLoss()   // The loss function
07.   If $Loss$ < 0.01   // Convergence
08.   Break
09. End
Output: P, Q

4.2 Individual Score Prediction Based on the Bayesian Model

The bayesian personalized ranking model predicts and enhances the relationship between users and items based on users' scores. Building upon the relationship between individuals and groups, this paper extends the bayesian personalized ranking algorithm to group-level and subsequently applies the matrix factorization strategy mentioned above.

This part adopts a bayesian model based on matrix factorization, and the process is shown in Algorithm 2.

Algorithm 2. Bayesian Personalized Sort Algorithm Based on Matrix Factorization

---

Input: Training set *train_txt*, learning rate α
Output: Personalized recommendation list L
01. $R = P * Q$
02. nP, nQ = matrix_factorization($P, Q, R, d, steps$)   // Latent factor dimension d
03. While ($Loss < minLoss$)   // Iterative update
04. $User$ = int(np.random.choice($train\_user$, 1))   // Pick a user at random
05. $ItemI$ = int(np.random.choice($visited$, 1))   // Pick an item I that user likes
06. $ItemJ$ = int(np.random.choice(all_item, 1))
07.   While $itemJ$ is visited   // If hit $itemJ$, choose another
08. $ItemJ$ = int(np.random.choice(all_item, 1))
09.   For $r = 1$ to R
10. $Factor$ = 1.0/($first-last$)
11.   End For
12. $Loss$ = countLoss()
13. $\{Y, w\}$ = updateWeights($Loss, a$)
14. End While   // Iteration finished
15. For $i = 1$ to N
16.   Array[i] = topk($R[i]$)
17. End For
Output: L

4.3 Positive Group Recommendation Strategy

When the group is a positive group, there are five recommendation strategies, which are the positive group recommendation strategy based on rating summation, the positive group recommendation strategy based on borda counting, the positive group recommendation strategy based on minimum tolerance, the positive group recommendation strategy based on intolerance average satisfaction, and the positive group recommendation strategy based on maximum satisfaction.

(1) The Positive group recommendation strategy based on rating summation.

In this strategy, the scores of all users in the item are summed and treated as the scores of the item in the matrix formed by the personalized recommendation list derived from matrix factorization. This results in the score list for this item, as defined as (5). Where Rij is the user i's rating of item j.

$$G_i = \sum_{i=1}^{n} R_{ij} \tag{5}$$

(2) The positive group recommendation strategy based on borda counting.

The borda counting method, originally an election system, resets the item score based on the user's score. Starting from 0, the lowest possible score, points are incrementally added in a consecutive fashion. If the item scores are identical, resetting the score becomes inconvenient. The item's final score is derived by adding the incremental points, thus obtaining the ultimate score.

(3) The positive group recommendation strategy based on minimum tolerance.

The strategy selects the highest score among all users in the item column of the matrix formed by the personalized recommendation list derived from matrix factorization, thereby obtaining the score list for this item. This strategy can enhance overall satisfaction, as demonstrated in (6).

$$G_i = \max_c r_{ij} \tag{6}$$

(4) The positive group recommendation strategy based on intolerance average satisfaction.

The strategy establishes a threshold value. If a user's score for an item is below this threshold, the score is discarded, and the remaining scores are summed. This approach can effectively enhance the overall satisfaction of the group.

(5) The positive group recommendation strategy based on maximum satisfaction.

In this strategy, the lowest score among all users in the item column is assigned as the score of the item in the matrix formed by the personalized recommendation list derived from matrix factorization. This results in the score list for this item, as illustrated in (7).

$$G_i = \min_c r_{ij} \tag{7}$$

4.4   Conformity Group Recommendation Strategy

When the group is a conformity group, there are three recommendation strategies, which are the conformity group recommendation strategy based on weighted rating summation, the conformity group recommendation strategy based on weighted tolerance, and the conformity group recommendation strategy based on weighted satisfaction.

(1) The conformity group recommendation strategy based on weighted rating summation.

The strategy is a weighted algorithm for the group with more passive users. Assuming that the positive user score weight value is α and the passive user score value is $1-\alpha$, the users in the item column give different weights according to the user attribute score, and then sum their scores to get the score list of this item (defined as (8)).

$$G_i = \alpha \sum_{i=1}^{n} R_{ij} + (1 - \alpha) \sum_{i=1}^{m} R_{ij} \tag{8}$$

(2) The conformity group recommendation strategy based on weighted tolerance.

This strategy is a weighted algorithm for the group with more passive users. Assuming that the positive user's score weight value is α and the passive user's weight value is $1-\alpha$, the user in the item column selects the lowest score according to the user's score multiplied by the weight, and obtains the score list of this item (defined as (9)).

$$G_i = \min_c(\alpha \cdot r_{ij}, (1 - \alpha) \cdot r_{ij}) \tag{9}$$

(3) The conformity group recommendation strategy based on weighted satisfaction.

This strategy is a weighted algorithm for the group with more passive users. Assuming that the positive user's score weight value is α and the passive user's weight value is $1-\alpha$, the user in the item column selects the lowest score according to the user's score multiplied by the weight, and obtains the score list of this item (defined as (10)).

$$G_i = \max_c(\alpha \cdot r_{ij}, (1 - \alpha) \cdot r_{ij}) \qquad (10)$$

5   Experiments

5.1   Dataset

In this paper, the social network dataset CiteULike is used as the experimental dataset, which is shown in Table 1. We selected some relatively active groups and users, resulting in an experimental dataset consisting of 23,031 (group-user-item) sets, including 8,679 papers. Given that the dataset itself contains triples, it was relatively straightforward to conduct experiments.

Table 1. Experimental datasets

Dataset	Number of groups	Number of users	Number of items	Number of Group-User-Item interactions
CiteULike	239	6040	3952	1000209

5.2   Experimental Evaluation Index

This paper employs three evaluation indices, which are Mean Average Preference (MAP), Recall, and MRR (Mean Reciprocal Rank), respectively.

MAP: The average accuracy mean, as its name suggests, indicates the accuracy of the recommended results, which is represented by the average value of the results. Equation (11) illustrates this concept.

$$MAP = \frac{1}{G}\sum_1^G(\frac{1}{I}\sum_1^I \alpha_n^c \cdot \mathbb{I}_n) \qquad (11)$$

In the Eq. (11), G is the number of test groups in experimental data, I is the number of users in a group, $\mathbb{I}_n$ represents the respective hit rates of the top n items from high to low, generally n value is 10, and α_n^c is a variable with a value of 1 or 0. If the top n items appear in the test group, the α_n^c value of this group is 1, and if not, it is 0.

Recall: The index reflects the number of correctly predicted positive examples in the experiment. The recall serves as an indicator to assess the performance of the recommendation system. Equation (12) represents this concept.

$$Recall = \frac{N^{TRUE}}{N} \tag{12}$$

In the Eq. (12), N is the number of items hit in the test, and N^{TRUE} is the number of items hit predicted by the recommended algorithm.

MRR: The mean reciprocal ranking (MRR) evaluates the strengths and weaknesses of the recommendation system, reflecting whether specific items should be included in the recommendation list. Equation (13) represents this concept.

$$MRR = \frac{1}{G}\sum_1^G \frac{1}{p_i} \tag{13}$$

In the Eq. (13), G represents the number of groups, that is, the number of recommendations, and p_i is the ranking of the items visited by the user in the formed recommendation list, that is, if the recommendation list does not hit this item, the value of p_i is infinite, and $\frac{1}{p_i}$ of this item is 0.

5.3 Group Classification Test

Through the learning of W^I and W^g, if individual users in most groups demonstrate a higher level of independent choice rather than being significantly influenced by the group, then these users are considered positive. On the contrary, users who are more influenced by the group are considered conformity. Table 2 shows the results of the group classification.

Table 2. Group classification results

Dataset	Number of groups	Number of positive groups	Number of conformity groups
CiteULike	6040	3854	2186

5.4 Group Recommendation Performance Test

In the Sect. 5.3, the classification of users and groups through group recommendation is implemented. Then, we adopt different group recommendation strategies based on the specific conditions of each group. In this section, we select several traditional recommendation models to compare with our presented DGRM model. These baseline models include the Collaborative Filtering (User-CF and Item-CF), and Bayesian Personalized Ranking (BPR). The two recommendation strategies used in this paper are

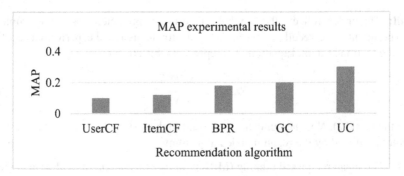

Fig. 2. MAP experimental results

the group recommendation strategy from the group's perspective (GC) and the group recommendation strategy from the user's perspective (UC).

The performance test results of MAP are shown in Fig. 2. The group recommendation strategy proposed in this paper has higher recommendation accuracy than the two algorithms of collaborative filtering (UserCF and ItemCF), and the bayesian personalized ranking (BPR) algorithm is slightly lower than the two group recommendation strategies. The accuracy of the group recommendation strategy from the user's perspective is higher than the accuracy of the group recommendation strategy from the group's perspective.

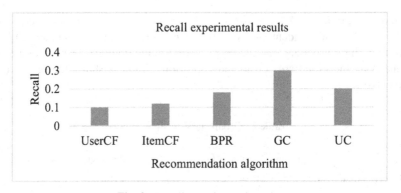

Fig. 3. Recall experimental results

The performance test results of recall are shown in Fig. 3. The recall of the user-based collaborative filtering algorithm (UserCF) is the lowest, and the gap is relatively obvious. The group recommendation strategy based on the user's perspective is similar to the BPR algorithm, and the highest is the group recommendation strategy based on the group's perspective.

The performance test results of MRR are shown in Fig. 4. The GC algorithm performs less well than the BPR algorithm, but the user-based group recommendation algorithm is still far ahead in the metrics, so it seems that the UC algorithm can better reflect the order and preferences of users who really visit the items.

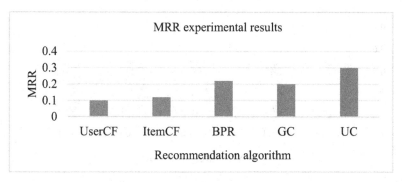

Fig. 4. MRR experimental results

5.5 Sensitivity Analysis

As observed from Fig. 3, the recall index of the group recommendation strategy proposed (GC) in this paper is lower than that of the BPR model. When most indices are categorized into the above groups, the classification criteria for positive and conformity groups are whether the proportion of positive users in the group reaches 50%. Variations in group classification conditions may also significantly impact the classification results. Table 3 shows the results under different conditions, that is, various proportions of positive users in the group, with the sensitivity analysis of classification results conducted on experimental datasets. As the conditions for positive group classification improve, the number of positive groups in the experimental data set decreases. Figure 5 depicts the histogram analysis of the sensitivity experimental results from the CiteULike dataset. The results for group classification and group recommendation remain relatively stable between 0.5 and 0.6 positive users. If the number of positive users surpasses 0.7, the impact of group recommendation experiences a substantial decline. Under the condition where positive users account for 0.55 within the group, the recommendation effectiveness for the experimental data set is at its best.

Table 3. Group classification under different conditions

Group classification	>=0.5	>=0.55	>=0.6	>=0.7
Number of positive groups	168	163	151	134
Number of conformity groups	71	76	88	105

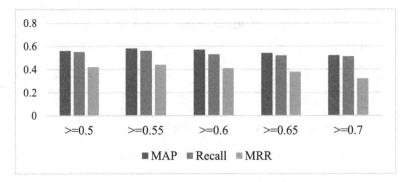

Fig. 5. Sensitivity test results

6 Conclusions

In this paper, we present the Diversified Group Recommendation Model for social network (DGRM). This model considers two aspects of social choice and social influence, models the diversity of groups, and adopts different group recommendation strategies for different groups, which can better meet the diverse needs of users and groups. We propose a group recommendation strategy based on score fusion, which can better meet the diverse needs of users and groups. Finally, the experimental results demonstrate the effectiveness and efficiency of our DGRM model.

Acknowledgement. This work was supported by the Science and Technology Program Major Project of Liaoning Province of China under Grant No. 2022JH1/10400009, the Natural Science Foundation of Liaoning Province of China under Grant No. 2022-MS-171, the Social Science Program Foundation of Liaoning Province of China under Grant No. L23BJY018, the National Natural Science Foundation of China under Grant No. 62072084.

References

1. Sun, C.C.: Research on Group Discovery and Recommendation for Social Networks. Nanjing University of Posts and Telecommunications (2019)
2. Pan, Y.: Research on Group Recommendation Based on Fusion Target Preference. Nanjing University of Finance and Economics (2016)
3. Zhang, Y.J., Du, Y.L., Meng, X.W.: Group recommendation system and its application research. J. Comput. Sci. **39**(04), 745–764 (2016)
4. Deng, A.L., Zhu, Y.Y., Shi, B.L.: Collaborative filtering recommendation algorithm based on item score prediction. Acta Softw. Sinica. **09**, 1621–1628 (2003)
5. Jameson, A., Baldes, S., Kleinbauer T.: Two methods for enhancing mutual warning in a group recommender system. Adv. Vis. Interfaces (2004)
6. Liu, S.Y., Chen, T.: Restaurant group recommendation model and system design based on social network. Innov. Appl. Sci. Technol. **8**, 27–30 (2019)
7. He, J., Liu, Y.Z., Wang, J.K.: Research on social network community classification and group recommendation strategy based on social choice and social influence. Mod. Intell. **38**(01), 92–99 (2018)

8. Salam, M., McCarthy, K., Smyth, B.: Generating recommendations for consultation negotiation in group personalization services. Pers. Ubiquitous Comput. **16**(5) (2012)
9. Guo, Z.W.: Research on group recommendation based on multi-feature fusion in social networks. Chongqing University (2018)
10. Qi, J., Liu, Y., Liu, Y.X., Hu, M.Z., Le, H.F.: Collaborative filtering recommendation method based on tag research. J. Beijing Union Univ. **35**(02), 47–52 (2021)
11. Rendle, S., Thieme, L.S.: Pairwise interaction tensor factorization for personalized tag recommendation. Web Search and Data Mining (2010)
12. Zhao, H.F., Wang, X.J.: Bi-group Bayesian personalized ranking from implicit feedback. In: Proceedings of 2019 2nd International Conference on Computer Science and Software Engineering (CCSE2019), pp. 42–46 (2019)
13. Koren, Y, Bell, R, Volinsky, C.: Matrix factorization techniques for recommender systems. Computer **42**(8) (2009)

PSL-Based Interpretable Generation Model for Recommendation

Dong Li[1], Binghao Han[1], Ming Wan[1]([✉]), Yuqian Gong[2], Yue Kou[2], and Hairong Liao[1]

[1] School of Information, Liaoning University, Shenyang 110036, China
dongli@lnu.edu.cn
[2] School of Computer Science and Engineering, Northeastern University, Shenyang 110819, China

Abstract. Nowadays, recommendation systems have been widely used in various aspects such as news, movies, music, videos, academia, and many more. The advent of personalized recommendation systems has significantly enhanced the efficiency of users' access to information and also improved their overall experience. As an essential component of the recommendation system research field, interpretable recommendations emphasize the need to provide users with recommended results along with the rationale behind them. Unlike traditional recommendation systems, interpretable systems can not only enhance system transparency but also increase user trust and acceptance, the likelihood of users choosing the recommended product, and overall satisfaction. However, most existing interpretable recommendation systems depend on user similarity, item similarity, scoring data, or review a single aspect of the data to produce an interpretation, which makes it challenging to create trustworthy interpretable due to the limited use of these factors and intelligent reasoning. To solve this problem, we propose the Probabilistic Soft Logic (PSL)-based Interpretable Generation Model for Recommendation (called PIGM). Unlike traditional interpretable recommendation models, our PIGM takes into account user similarity, item similarity, and scoring data, employing PSL to model these factors and utilizing intelligent reasoning to generate interpretations. Firstly, based on matrix decomposition to predict the user's score of the item. Secondly, the data is partitioned into observation dataset, target dataset and fact dataset. Thirdly, predicates are defined according to the data and the required results, and rules are defined according to the defined predicates, fact logic and relevant knowledge generated by recommendation system interpretation. Then, the weights of the rules are learned based on the maximum likelihood estimation. Finally, based on PSL reasoning and according to the defined rules, generate a recommendation list and corresponding interpretation for the user. The experimental results demonstrate the effectiveness of our proposed PIGM model on the real dataset.

Keywords: Interpretable Recommendation · Probabilistic Soft Logic (PSL) · Interpretation Generation · PSL Reasoning

© The Author(s), under exclusive license to Springer Nature Singapore Pte Ltd. 2024
X. Song et al. (Eds.): APWeb-WAIM Workshops 2023, CCIS 2094, pp. 52–64, 2024.
https://doi.org/10.1007/978-981-97-2991-3_5

1 Introduction

According to the user's personal information and purchase preference, the interpretable recommendation system recommends the items that the user may be interested in and explains recommending the items. To effectively alleviate the impact of information overload on users, interpretable recommendations can assist users in making better choices, thereby enhancing the transparency and satisfaction of the recommendation process. At the same time, the existing interpretable recommendation methods also have some shortcomings. Through an analysis of its research status, it can be observed that the current interpretable recommendation methods have several shortcomings.

(1) Most of the traditional personalized recommendation methods focus on the accuracy of the recommendation results, ignoring the interpretation of the recommendation results. The interpretable recommendations can not only make personalized recommendations to users but also produce corresponding interpretations so that users can understand why to recommend specific items and improve the credibility of recommendation.

(2) Existing interpretable recommendation methods are user-based and template-based, which have the disadvantages of high repetition rate of interpretation sentences and lack of individuation. Not only is there a need to better model users and improve recommendations for them, but there is also a growing need for interpretability.

There are many available datasets, including user feedback and comments usually written by users. These datasets contribute to the research and development of new, data-driven technologies for interpretable recommendation systems. The interpretable recommendation can provide explanations and item suggestions to enable users to understand why a particular item is suggested, thereby making better decisions, improving the overall transparency of the system, and enhancing user-perceived credibility.

In this paper, we propose the Probabilistic Soft Logic (PSL)-based Interpretable Generation Model for Recommendation (called PIGM). More specifically, we make the following contributions.

(1) The PIGM model is proposed. Unlike traditional interpretable recommendation models, our PIGM takes into account user similarity, item similarity, and scoring data, employing PSL to model these factors and utilizing intelligent reasoning to generate interpretations.

(2) Based on matrix decomposition, the user's score for the item is predicted. Moreover, the predicates are defined according to the data and the required results, and rules are defined according to the defined predicates, fact logic and relevant knowledge generated by the recommendation system interpretation. Then, the weights of the rules are learned based on maximum likelihood estimation.

(3) Based on PSL reasoning and according to the defined rules, generate a recommendation list and corresponding interpretation for the user.

(4) The experimental results demonstrate the effectiveness of our proposed PIGM model on the real dataset.

2 Related Work

With the rapid development of e-commerce platforms, users are faced with massive information. The recommendation system aims to provide users with items and effective services that are more in line with their preferences by recording users' historical behaviors and learning users' preferences. The recommendation system can not need users to provide clear item preferences and needs, but by analyzing the relevant historical behavior data information of users and items, model users and items, to actively recommend items that users may be interested in. In recent years, more and more people focus on generating explanatory recommendations, because providing interpretations to users can help users make better and faster decisions.

In the traditional recommendation model, the algorithm focuses on the accuracy of recommendation results. For example, the user-based collaborative filtering/item-based collaborative filtering model [1, 2] and CF model (UserCF/ItemCF), whose principle is to construct a user-item co-occurrence matrix according to the historical actions of users, and make sequential recommendations based on similar users or similar items. Matrix decomposition model [3], whose principle is to decompose the co-occurrence matrix in CF into user matrix and item matrix, and make sequential recommendations by using the inner product of the user implicit vector and item implicit vector. Logistic regression model [4], the principle of which is to input the information of users, articles, and context into the logistic regression model to get CTR (Click Through Rate), and then sort and recommend according to CTR. The principle of the factorization machine model is to add a second-order feature crossover based on LR (Logistic Regression), train a hidden vector for each one-dimensional feature, and get the weight of the crossover feature by inner product operation between hidden vectors.

In the interpretable recommendation model, not only the personalized recommendation can be made to users, but also the corresponding interpretation can be generated. For example, in the interpretable recommendation model based on users and items [5], by analyzing the historical item data of target users and other users, when an item is recommended to target users, the interpretation given is that users with similar preferences have a high evaluation of the item. Feature-based interpretable recommendation model [6], which explains that the features in the recommendation items match the preferences of the target users. Topic-based interpretable recommendation model [7], e-commerce websites have widely available text comments, and the hidden topics learned from comments can understand the hidden factors in the potential factor model and improve the accuracy of score prediction. It can further explain the score given by users to a certain item, and then get the personalized preference of each user for different items, to make better recommendations. Interpretable recommendation model based on deep learning [8, 9], which covers a wide range of deep learning technologies, including CNN (Convolutional Neural Network) [10], RNN (Recurrent Neural Network) [11], memory network, and so on. Based on the natural language generation model, the recommendation system can automatically generate interpretation statements. The model connects user ratings to the input component as ancillary information, so that the model can generate comments based on predicted ratings (emotions).

Maximum likelihood estimation (MLE) [12] is an important and universal method to find estimators. It is a method to estimate the parameters of the probability distribution model based on the known observation results (samples) and the given probability distribution model, and it is the most likely method to generate this known sample under this parameter. In other words, the optimal linear model maximizes the probability of samples X and y. Therefore, only one error can be added to the prediction $\varepsilon(i)$: $y(i) = hw(x(i)) - \varepsilon(i)$. Where $y(i)$ denotes the actual value of a sample and $hw(x(i))$ denotes the value predicted using the model. $\varepsilon(i)$ denotes Error. The errors of all samples are independent and identically distributed and obey Gaussian distribution (normal distribution) with a mean value of 0 and variance of a certain value.

Knowledge Graph (KG) [13] can express the rich entity relationship between users and items by complex network, dig out the deep hidden interaction relationship between users and items, and improve the accuracy of recommendation. The extracted fine-grained relationship between users and items is also helpful to generate more personalized explanations for recommended items and improve the interpretability of the recommendation system.

Probabilistic soft logic [14], a new probabilistic programming language, makes HL-MRFs (Hinge Loss Markov Random Fields) easy to define and apply to large relational datasets. This idea has been explored in other types of models, such as Markov logic networks of discrete MRF, relational dependent networks of dependent networks, and probabilistic relational models of Bayesian networks. We define PSL based on these previous methods and the relationship between hinge loss potential and logic clause. In addition to probability rules, PSL also provides syntax, which enables users to easily apply many common modeling techniques, such as domain and range constraints, blocking and canopy functions, and aggregating variables defined on other random variables. PSL allows HL-MRFs to be easily applied to a wide range of structured machine-learning problems by defining potential and constrained templates.

Maximum likelihood estimation and probabilistic soft logic are the foundation of the PSL-based Interpretation Generation Model (PIGM) proposed in this paper. Different from the traditional interpretable recommendation model, this model comprehensively considers user similarity, item similarity, and score data, and uses PSL to model these factors and intelligent reasoning to generate interpretation.

3 Model Overview

Probabilistic Soft Logic (PSL) is a collective probabilistic reasoning framework in relationship area. The PSL uses first-order logic rules as the template language for graphical models for random variables with soft truth values from intervals [0, 1]. Reasoning under this setting is a continuous optimization task, which can be effectively solved. In this paper, we propose the Probabilistic Soft Logic (PSL)-based Interpretable Generation Model for Recommendation (called PIGM). The framework of the PIGM model is shown in Fig. 1.

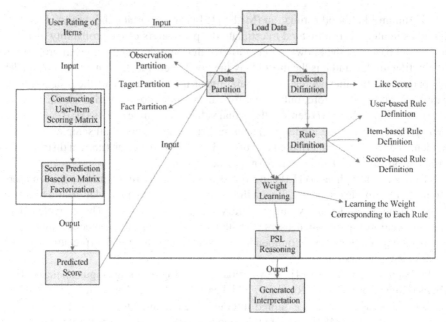

Fig. 1. PIGM framework

4 PIGM Model

4.1 Score Prediction Based on Matrix Factorization

Score prediction is based on matrix factorization, and the result without score is predicted by using score dataset. Matrix factorization is to decompose a matrix into the product of one or two matrices. For the user rating matrix in Table 1, it is recorded as $R_{m \times n}$. It can be decomposed into the product of two matrices, assuming that it is decomposed into two matrices $P_{m \times k}$ and $Q_{k \times n}$. We want the product of matrices $P_{m \times k}$ and $Q_{k \times n}$ to restore the original matrix $R_{m \times n}$. As shown in Eq. (1).

$$R_{m \times n} \approx P_{m \times k} \times Q_{k \times n} = R_{m \times n} \tag{1}$$

Where $P_{m \times k}$ represents the relationship between m users and k topics, while $Q_{k \times n}$ represents the relationship between k qualified topics and n users. We use the average of the errors between the original scoring matrix $R_{m \times n}$ and the newly constructed scoring matrix $R_{m \times n}$ as the loss function, as shown in Eq. (2)

$$e_{i,j}^2 = \left(r_{i,j} - \hat{r}_{i,j}\right) = \left(r_{i,j} - \sum^k p_{i,k} q_{k,j}\right) \tag{2}$$

Then solve the minimum value of the sum of losses of all not "-" terms min $=$ $loss \sum e_{i,j}^2$. by gradient descent method, first solve the negative gradient of the loss function, as shown in Eqs. (3) and (4).

$$\frac{\partial}{\partial q_{i,k}} e_{i,j}^2 = -2\left(r_{i,j} - \sum_{n=1}^k p_{i,k} q_{k,j}\right) q_{k,j} = -2e_{i,j} q_{k,j} \tag{3}$$

Table 1. User rating matrix

	M1	M2	M3	M4
P1	5	3	-	1
P2	4	-	-	1
P3	1	1	-	5
P4	1	-	-	4
P5	-	1	5	4

$$\frac{\partial}{\partial q_{k,j}}e_{i,j}^2 = -2\left(r_{i,j} - \sum_{n=1}^{k} p_{i,k}q_{k,j}\right)p_{i,k} = -2e_{i,j}q_{k,j} \tag{4}$$

Next, update the variables according to the direction of the negative gradient, as shown in Eqs. (5) and (6).

$$p_{i,k}' = p_{i,k} - \alpha\frac{\partial}{\partial p_{i,k}}e_{i,j}^2 = p_{i,k} + 2\alpha e_{i,j}q_{k,j} \tag{5}$$

$$q_{k,j}' = q_{k,j} - \alpha\frac{\partial}{\partial q_{k,j}}e_{i,j}^2 = q_{k,j} + 2\alpha e_{i,j}p_{i,k} \tag{6}$$

Via iteration, we know that the algorithm finally converges. Using the above process, two matrices $P_{m\times k}$ and $Q_{k\times n}$ can be obtained finally, so that the score of user i on commodity j can be inferred.

4.2 PSL Rule Definition

PSL supports logical rules and arithmetic rules. Each of these types of rules supports weights and squares, and the general form of the rule definition is shown in Eq. (7).

$$w: P(A, B)\&Q(B, C) \rightarrow R(A, C) \tag{7}$$

The first part w of the rule is a floating-point value that specifies the weight of the rule. In this example, P, Q, and R are predicates, and A, B, and C are items. Logical rules consist of the rule "body" and rule "header". The body of the rule appears before the logical meaning. A body may have one or more predicates joined together with a logical conjunction operator, and a header may have one or more predicates disjunctive with a logical disjunctive operator. Predicates that appear in the body and header can be any combination of open predicate types and closed predicate types.

4.3 Rule Weight Learning Based on Maximum Likelihood Estimation

Each rule must be weighted or unweighted. Unweighted rules are also called constraints because they are strictly enforced. The weights of rules can be learned by maximum likelihood estimation. The gradient of logarithmic likelihood value relative to weight is shown in Eq. (8).

$$\frac{\partial}{\partial \lambda_i} \log f(I) = -\sum_{r \in R_i} (d_r(I))^p + E\left[\sum_{r \in R_i} (d_r(I))^p\right] \qquad (8)$$

Among them, R_i is a basic rule set parameterized by weight λ_i. It is difficult to calculate the expectation $E[\sum_{r \in R_i} (d_r(I))^p]$, so a common approximation is used $\sum_{r \in R_i} (d_r(I^*))^p$. This is the most likely interpretation for a given current weight. In addition, the new PSL weight learning method is an active research field.

$$SAME(M1, M2)\&\&LIKE(P1, M1) \rightarrow LIKE(P1, M2) \qquad (9)$$

Equation (9) is a rule. Assuming that the probabilities of $SAME(M1, M2)$, $LIKE(P1, M1)$, and $LIKE(P1, M2)$ under this rule are 1, 0.9, and 0.3, respectively, then, we can make a conclusion:

(1) Find the probability that the rule body (the condition on the left side of the rule) holds I $(rbody) = \max \{0, 1 + 0.9 - 1\} = 0.9$.
(2) Calculate the rule loss (the distance between the rule header and the rule body) d_r $(I) = \max \{0, 0.9 - 0.3\} = 0.6$.

For each instance under each rule, the above calculation process is adopted to calculate the rule loss. Our goal is to minimize the sum of these rule losses, that is, to maximize the probability density function $f(I)$, which also requires the maximum of $f(I)$ under what combination of weights. This is a probability optimization problem, which can be done by maximum likelihood estimation, and the weights are estimated.

4.4 PSL-Based Interpretable Generation

The first common reasoning task in PSL is to find the most likely interpretation of a given evidence, that is, to extend a given partial interpretation. This means that the density function $f(I)$ in the maximization equation (Eq. 8), which is equivalent to the sum of the minimization exponents, is subject to evidence and equality and inequality constraints. As shown by Brow et al., the constrained optimization problem can be transformed into a second-order cone program (SOCP). SOCP can be solved in time O $(n^{3.5})$, where n is the number of relevant rule bases, that is, those satisfied with non-zero distance. To avoid manipulating irrelevant rules, PSL uses an iterative method before constructing SOCP, which determines the relevant rule set based on the truth value of the evidence atom and the current truth value of the non-evidence atom. Initially, the truth value for non-evidential atoms is 0. After building and solving SOCP, update the relevant rule set according to the current MPE (Most Probable Explanation) interpretation. This process will be repeated until no more rules are activated.

The MPE reasoning based on consensus optimization can achieve linear scalability, but its accuracy is only slightly lower than the standard cubic time SOCP solver used in the above methods. Consensus optimization decomposes the optimization problem into independent small problems linked by additional constraints. In PSL, separate sub-questions are created for each basic rule. Each such subproblem uses its copies of local text and introduces constraints that equate the true values of these local copies to the true values of the corresponding original text.

In PSL, partitions are used to organize data. Partitions are just containers of data, but we use them to keep specific blocks of data together or separately. For example, if we are running the results evaluation model, we must ensure that our test partition is not used in training.

PSL users typically organize their data in at least three different partitions.

(1) Observation Partition is the observation partition. In this partition, we put the actual observation data. In this example, we put all the observations in the observation section about the user's rating of the magazine.
(2) Target Partition is the target partition. In this partition, we place atoms of the values we want to infer. For example, if we want to infer the probability that $P1$ likes magazine $M1$, we will put atom *LIKE* $(P1, M1)$ into the target partition.
(3) Truth Partition is the fact partition. in this partition, we put the actual value data, but not included it in our known observation data, to evaluate the reasoning effect of the model. For example, if we knew that $P1$ liked magazine $M1$, we would put the *LIKE* $(P1, M1)$ fact partition into the truth value 1.

4.5   Interpretation Generation Algorithm

The interpretation generation algorithm based on PSL reasoning needs to partition data, define predicates and define rules, all of which need to be defined according to datasets. After the weight learning, this paper is based on the maximum likelihood estimation of the rule weight learning, then runs the reasoning to explain the generation. After that, the output is carried out, and the output results are evaluated.

Algorithm 1 is the description of the interpretation generation algorithm based on PSL reasoning. The input is to score the item for known users, and then the interpretation statement is finally generated through predicate definition, rule definition, weight learning, and running reasoning.

Algorithm 1. Interpretation Generation Algorithm Based on PSL Reasoning

Input: The user's rating of the item
Output: The interpreting statements
```
01.  model ←MODEL_NAME = 'simple-acquaintances'   // Define the PSL model
02.  predicate ← 'LIKE' 'SCORE'           // Get the predicate definition
03.  get_predicates(self)
04.    name = predicate.name()
05.    self.predicates[predicate.name()] = predicate
06.  add_rules(model)                      // Add predicate definitions to the model
07.  rules ← defined by rule
08.  get_rules(self)
09.    self.rules.append (rule)            // Get the rule definition
10.  add_rules(model)                      // Add rule definitions to the model
11.  for i in range(len(self._rules))      // Weight learning
12.    if (self.rules[i].weighted())
13.      self.rules[i].set_weight(new_weights[i])
14.  results ← infer(model)                // PSL reasoning
15.    add_data(model)
16.    return model.infer()
17.  write_results(results, model)         // Generate interpretation
```

5 Experiments

5.1 Dataset

This paper uses only rating data in magazine subscriptions in the Amazon review dataset, and the dataset obtained on the network only includes entry, user, rating, and timestamp tuples. Because the PIGM model implemented in this paper does not need a timestamp, we only keep the first three tuples. The items and users given in this dataset are represented by long codes. For the convenience of use, we sort the items and users from 1. There are more than 80,000 pieces of data given in the dataset. To reduce the workload of this paper, we choose the scoring data of the first 10,000 users, which means 12,351 scoring data. We will work based on these data. As shown in Table 2 and Table 3.

Table 2. Samples of experimental data

User serial number	Magazine serial number	Score
1	248	1
1	343	4
…	…	…
501	2776	4
…	…	…

Table 3. Statistic of experimental data

Dataset	Magazine Subscriptions
Users	10000
Magazine	1474
Score	12351

5.2 Accuracy Analysis of Score Prediction

After the parameters are determined, it is necessary to train and output based on PSL interpretation, because the training result is that each product corresponds to a user. The results are listed in a two-dimensional matrix, as shown in Table 4.

Table 4. Interpretation the two-dimensional matrix

Favorite probability rate	M1	M2	M3	M4	M5	...
P1	0.68	0.93	0.12	0.53	0.22	...
P2	0.17	0.36	0.98	0.45	0.99	...
P3	0.58	0.62	0.42	0.33	0.71	...
P4	0.28	0.14	0.85	0.41	0.88	...
P5	0.01	0.90	0.22	0.56	0.98	...
...	...	...	...	...	...	...

The first column of the Table 4 represents all users, the first row represents all magazines, and the middle table is the predicted liking probability of each user for each magazine (0–1). The larger the value, the more likely it is to like it.

We compare the resulting target data with the real score data. As shown in Table 5, the first column is the user number, the second column is the magazine number, the third column is the probability that the user may like the corresponding item, and the fourth column is the real data retained in our experiment. The real data is defined as "1" for like and "0" for dislike. It can be seen that if the data in the third column is close to 1, it is also shown as 1 in the real data, that is, the user likes the item. Conversely, if the data in the third column is close to 0, it will be 0 in the real data, which means that the user doesn't like the item. After comparison, the experimental results are more accurate, and the accuracy of the model can be determined.

We calculate the RMSE and MAE of this model and compare them with the values of traditional SVD++ (Singular Value Decomposition). The results are shown in Table 6.

According to Table 6, it can be seen that the model algorithm proposed in this paper is relatively low in RMSE and MAE evaluation index compared with the SVD++ algorithm, which shows that the overall recommendation performance of this model is better than the SVD++ algorithm.

Table 5. Comparison of predicted data and real data

User ID	Magazine ID	Probability of possible liking	Real data
0	1	0.999197602	1
0	7	0.999079466	1
0	15	0.99852109	1
0	18	0.998974383	1
0	22	0.006305193	0
1	2	0.999777891	1
1	11	0.003111308	0
1	13	0.999189198	1
1	17	0.490572959	0
2	6	0.997023821	1
2	11	0.00220674	0
3	7	0.997023821	1
3	10	0.996961296	1
3	12	0.993002892	1
3	21	0.997935474	1
4	0	0.998141587	1
4	15	0.003479955	0
5	2	0.999514103	1
5	7	0.429610491	1
5	9	0.496092409	1
5	12	0.99866122	1

Table 6. Performance comparison

	PEGM	SVD++
RMSE	0.9018	0.9121
MAE	0.7368	0.7411

5.3 Interpretable Generation Analysis

According to the results in Table 5, it indicates the user's liking probability of magazines predicted by this model. The system can predict that the results predicted by this model are more accurate. We recommend magazines with higher probability to users, which can increase the purchase probability of users.

Interpretable generate based on PSL, and users can see the reasons for recommending products more clearly by looking at these interpretable. A sample of PSL-based interpretable generation is shown in Table 7.

Table 7. A sample of PSL-based interpretable generation

Users	Recommended merchandise	Similar users	Similar items	Score	Generated interpretation
1	2	2			Recommendations based on user 2 favorite products similar to your hobbies
2	6		17		Recommendations based on similar items you previously purchased 17
3	10		11	5	According to your previous recommendation of similar item 11 with a high score
4	15	14	6	5	Recommendation based on users with similar hobbies 14 similar item with higher scores 6

The PIGM can not only be generated based on similar users, such as the first line, but also integrate various factors, such as the fourth line, which integrates similar users, similar products and historical scores. This comprehensive multi-angle recommendation can be more similar to the real thoughts of users.

6 Conclusion

In this paper, we propose the Probabilistic Soft Logic (PSL)-based Interpretable Generation Model for Recommendation (called PIGM), which takes into account user similarity, project similarity, and scoring data, employing PSL to model these factors and utilizing intelligent reasoning to generate interpretations. Firstly, based on matrix decomposition to predict the user's score of the item. Secondly, predicates are defined according to the data and the required results, and rules are defined according to the defined predicates, fact logic and relevant knowledge generated by recommendation system interpretation. Then, based on PSL reasoning and according to the defined rules. Generate a recommendation list and corresponding interpretation for the user. Finally, the experimental results demonstrate the effectiveness of our proposed PIGM model.

Acknowledgement. This work was supported by the Science and Technology Program Major Project of Liaoning Province of China under Grant No. 2022JH1/10400009, the Natural Science

Foundation of Liaoning Province of China under Grant No. 2022-MS-171, the Social Science Program Foundation of Liaoning Province of China under Grant No. L23BJY018, the National Natural Science Foundation of China under Grant No. 62072084.

References

1. Lu, Q.: Research and implementation of recommendation system based on collaborative filtering model and argot semantic model. Hunan University (2013)
2. Liang, Q.: Research and Application of Recommendation Model Based on Collaborative Filtering. Beijing University of Posts and Telecommunications (2016)
3. Song, R.X., Li, G.Y.: Application of improved matrix decomposition model in recommendation system. Chinese Academy of Sciences (2019)
4. Wang, J.C., Guo, Z.G.: Logistic regression model: method and application. Higher Education Press (2001)
5. Zhang, Y., Chen, X.: Interpretable generation a survey and new perspectives. Found. Trends Inf. Retr. 14(1), 1–101 (2020)
6. Hou, Y., Yang, N., Wu, Y., et al.: Interpretable generation with a fusion of aspect information. In: World Wide Web Conference, vol. 22, no. 1, pp. 221–240 (2019)
7. Ren, Z., Liang, S., Li, P., et al.: Social collaborative viewpoint regression with interpretable generations. In: Proceedings of the 10th ACM International Conference on Web Search and Data Mining (WSDM), pp. 485–494 (2017)
8. Koren, Y., Bell, R., Volinsky, C.: Matrix factorization techniques for recommender systems. Computer 42(8), 30–37 (2009)
9. Gao, J., Wang, X., Wang, Y., et al.: Interpretable generation through attentive multi-view learning. In: Proceedings of the AAAI Conference on Artificial Intelligence, pp.3622–3629 (2019)
10. Yann, L.C., Leon, B., Bengio, Y., et al.: Gradient-based learning applied to document recognition. Proc. IEEE 83(11), 2278–2234 (1998)
11. Liu, J.W., Song, Z.Y.: Review of recurrent neural networks. Control Decis.-Mak. 37(11), 2753–2768 (2022)
12. Li, L., Zhang, Y., Chen, L.: Generate neural template interpretable for recommendation. In: Proceedings of the 29th ACM International Conference on Information and Knowledge Management (CIKM), pp. 755–764 (2020)
13. Angelika, K., Stephen, H., Matthias, B., Bert, H., Lise, G.: A short introduction to probabilistic soft logic (2013)
14. Zhu, D.L., Wen, Y., Wan, Z.C.: Review of recommendation system based on knowledge map. Data Anal. Knowl. Discov. 5(12), 1–13 (2021)

Personal Credit Data Sharing Scheme Based on Blockchain and Access Control

Jie Feng, Xiaoguang Li$^{(\boxtimes)}$, and Xiaoli Li

School of Information, Liaoning University, Shenyang 110036, China
xgli@lnu.edu.cn

Abstract. Personal credit plays a vital role in the modern economy and society. However, the traditional centralized credit model suffers from numerous issues, including privacy breaches, data misuse, and unclear data ownership. Moreover, this model lacks an efficient sharing mechanism, resulting in data dispersion and low utilization rates. To solve these problems, this paper proposes a personal credit data security sharing scheme based on multi-chain collaboration. In this scheme, credit data is stored in IPFS, and data summaries and information are uploaded to the master chain to ensure data integrity and consistency. Additionally, access control and key management operations are transferred from the master chain to reduce pressure and enable capacity expansion. Furthermore, to safeguard the security and privacy of data sharing, the paper designs a more efficient access control model called PT-ABAC, based on attribute-based access control (ABAC) and capability-based access control (CapBAC). The verification of static attributes is replaced by granting user permission token, and the zero-knowledge proof algorithm is used to protect the privacy of user attributes. The model realizes the fine-grained access control of data, and effectively improves the efficiency of data access. The analysis shows that the credit data sharing scheme in this paper is characterized by distributed, tamper-proof, traceable, secure and transparent data sharing, and fine-grained access control, which ensures the authenticity and immutability of the credit data, implements strict access control in credit data sharing, and clarifies the ownership and control of the credit data.

Keywords: Blockchain · Credit data sharing · Access control · Smart Contract

1 Introduction

Credit plays a significant role in modern economy and society. It serves as an essential credential for individuals to participate in social activities and is a crucial assessment indicator for banks, enterprises, and other institutions. However, the current personal credit system primarily relies on traditional centralized models, where credit bureaus act as central nodes collecting, processing, and storing data, and users retrieve the data from these bureaus [1]. Unfortunately, this model suffers from various drawbacks, including information asymmetry, inefficient data collection, a non-transparent information

Supported by the National Key Research and Development Program of China: (grant number: 2021YFF0901004)

environment, and limited credibility. Additionally, enterprises and government departments store their data in separate databases, resulting in data silos and hindering the development of a unified credit system [2]. Furthermore, credit data faces risks such as theft, tampering, and misuse. In 2022, numerous incidents of personal data leaks in banks occurred, leading to recurring cases of fraud and other illegal activities caused by the compromise of such information. The problems existing in the traditional credit investigation system can be summarized as follows. Firstly, the lack of effective sharing mechanism and data dispersion of personal credit data lead to low efficiency of collection and utilization of credit data. Secondly, the access of credit data does not impose strict access control, so there are risks of data theft, tampering and abuse [3]. The rights and interests of individuals are not protected, and the ownership of data is not clear.

The continuous advancement and expansion of blockchain technology have enabled its widespread application in various domains such as finance and the Internet of Things. Consequently, researchers have started exploring the utilization of blockchain in the field of credit collection. Blockchain technology possesses inherent characteristics like immutability, tamper resistance, and decentralized maintenance, making it an ideal distributed ledger for managing credit information [2]. For instance, Zhang employed blockchain in personal credit collection and developed a framework for a personal credit information sharing platform aimed at sharing credit blacklists [5]. Ding examined the personal credit system from a blockchain perspective and proposed the creation of a data sharing platform through a federation chain [1]. Ju et al. designed a decentralized lending and borrowing application using a blockchain-based big data credit platform to establish a credit platform that amalgamates data from multiple heterogeneous sources [6]. Chen et al. introduced a model for a blockchain storage and decentralized credit system that utilizes blockchain for storing sensitive user data, thereby enhancing data security [7]. Furthermore, Ta et al. analyzed the challenges associated with credit collection on Internet financial platforms and introduced a blockchain-based cross-platform credit data sharing model [2].

All the aforementioned solutions utilize blockchain technology to address the issue of data silos by designing credit data sharing or exchange platforms. Although the data is shared, there are still some problems such as privacy leakage and data abuse when the data is shared, and the information subject does not have the control over the data. To tackle these challenges, this paper proposes a solution by combining blockchain with access control technologies. Access control is a technology that ensures resource security and prevents unauthorized access according to predefined policies [8]. Blockchain offers benefits such as immutability and traceability, ensuring data security and integrity [9–11]. Smart contracts, on the other hand, fulfill various business requirements in diverse scenarios and find wide application in finance and other domains [12]. Through the combination of blockchain and access control, data security sharing can be realized. For instance, Zhang et al. proposed a cross-domain data sharing model that combines attribute-based access control (IDACM) with blockchain [13]. Chang et al. presented a blockchain-based data sharing model for multi-chain environments, incorporating hierarchical access control through smart contracts and security level configurations [14]. Ullah et al. introduced IoT Chain, an IoT-focused data sharing scheme that employs the ABAC model for fine-grained access control policies, as well as smart contracts for automating data access [15]. By Combining blockchain with access control, this paper aims to solve the problems of the credit system and establish a secure and controlled approach to credit data sharing.

The paper makes the following contributions: (1) A personal credit data security sharing scheme based on multi-chain collaboration is proposed. The scheme consists of a master chain, side chains and IPFS. The credit data is stored in IPFS, while the master chain stores the data summary information. Access control and key management operations are transferred to the side chains and anchored to the master chain based on smart contracts. The scheme ensures the authenticity and integrity of credit data, and clarifies the ownership of credit data while realizing data sharing. (2) Combining ABAC and CapBAC, the data access control model PT-ABAC is designed to realize fine-grained access control on data, Information subject can decide the access right of credit data by setting policy. PT-ABAC improves the efficiency of data access by setting permission tokens instead of static attribute verification, and uses zero-knowledge proof algorithm to protect user privacy. Access control is implemented by designing access control contracts and is responsible for storing and managing attributes and policies. PT-ABAC model provides secure and efficient access control for credit data.

2 Related Work

2.1 Access Control

Presently, the primary access control techniques are Role-Based Access Control (RBAC), Attribute-Based Access Control (ABAC), and Capability-Based Access Control (Cap-BAC). RBAC associates access to resources with roles, whereby users indirectly acquire privileges by obtaining specific roles [16]. ABAC operates based on subject and object attributes, enabling flexible, fine-grained, and dynamic control [8]. CapBAC implements access control using capabilities, granting functionality through tokens, and is relatively lightweight as it only needs to describe the subject's access rights [17]. These access control approaches are traditionally centralized. However, blockchain technology can effectively address access control challenges in distributed environments, and numerous access control schemes based on blockchain have been implemented.

2.2 zk-SNARK

zk-SNARK, which stands for Zero Knowledge Verifiable Non-Interactive Proof, is a cryptographic protocol utilized for implementing zero-knowledge proofs. The fundamental concept of zk-SNARK is to achieve the objective of proving the correctness of a specific statement to other parties while revealing no additional information. This is accomplished by transforming the proof process into a concise, non-interactive proof [18]. Some commonly employed zk-SNARK algorithms include groth16 [19], sonic, and others. In this paper, we adopt the groth16 algorithm due to its faster verification speed. The basic steps of zk-SNARK are as follows:

Setup(C) $\rightarrow$ (pk, vk): A trusted third party calls the setup algorithm to generate the public parameters needed for the "Prove" algorithm and the Verify algorithm. The

Setup algorithm takes as input the defined program C and generates pk, vk, where pk are the parameters of the proving side and vk are the parameters of the verifying side, the parameter's are used to facilitate both sides to generate and verify the short proofs.

Prove(pk, C, x, w) $\rightarrow$ π:The "Prove" algorithm takes the public key pk, program C, public input x, and private input w as its inputs. It then generates the proof π and transmits it to the verifier. π consists of 3 calculated elliptic curve points.

Verify(vk, π) $\rightarrow$ $result$: Verify proves the legitimacy of π by means of an equation verification, using vk and π as inputs with {true, false}.

In this paper, Zokrates tool is used to implement the construction of zk-SNARK algorithm. Zokrates is a zero-knowledge proof framework [21] to support off-chain computation and on-chain verification. The framework defines a specific language for generating arguments and computing proof π, and supports the automatic generation of Verify smart contracts.

3   Personal Credit Data Sharing Scheme Design

3.1   Overall Structure

This paper presents a personal credit data sharing scheme based on blockchain, aiming to address the issue of data silos. The scheme involves multiple participants who serve as network nodes, collectively maintaining the blockchain network. Instead of being stored solely within enterprises and organizations, credit data is returned to individuals and stored within the IPFS and blockchain. This approach enhances data security and mitigates the risks of data theft and misuse. Additionally, the paper introduces a data access control model named PT-ABAC to regulate data access.

As shown in Fig. 1, the scheme in this paper is mainly composed of two cooperative networks (blockchain network and IPFS network) and entities participating in credit data sharing. The blockchain is based on the multi-chain structure design, and two side chains are introduced on the basis of the master chain named AccesscontrolChain and KeyChain respectively, collectively referred to as service chain. Services such as access control and key management are transferred to the service chain to reduce the pressure on the master chain. The access control chain realizes the access control to the data and manages the attribute information, and the KeyChain is responsible for the key management and update; The anchor between the master chain and the service chain is realized through the smart contract protocol. IPFS is used to store credit data and generate a unique hash address CID for the data, which is updated on the master chain along with related information.

3.2   Entity

The entities involved in credit data sharing mainly include information subject DO, data generator, user DU, attribute authority AA and regulatory department.

(A) Information subject: The information subject is the credit data owner, and in the proposed scheme the information subject has the actual control over the credit data and decides the access to the credit data.

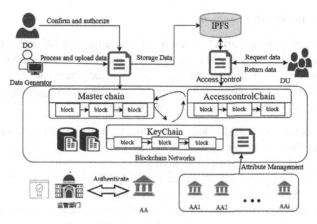

Fig. 1. Overall structure

(B) User: User refer to the user of credit data, including individuals, enterprises and other entities. Users request access to data through contracts.

(C) Data generator: Data generators refer to the third party enterprise or institution that interacts with the user to generate credit data.

(D) Attribute authority (*AA*): Enterprises and institutions with authority and high reputation in the industry are elected to manage attributes and supervise the platform.

(E) Regulatory authority: Government department or agency, mainly responsible for regulatory work, and as a certification authority CA.

3.3 Credit Data Sharing Process

(1) The credit data generator pre-processes the data and uploads it to the contract.

(2) Once the contract receives the data, it notifies the Information subject to confirm the data within a specified period of time.

(3) The information subject firstly confirms the credit data and restricts the access of users to the credit data by setting access policies.

(4) The confirmed credit data is encrypted using the AES algorithm and stored in IPFS. The hash address, *CID*, policy and other relevant information are updated on the master chain for sharing.

(5) The user sends a data access request to the access control contract. The access control contract determines whether the user has the right to access the data. The specific access control process is described in Sect. 4.

(6) If the user passes verification of the access control contract, the contract queries the data based on the *CID*, decrypts it, and returns it to the user. Otherwise, access is denied.

(7) The user's access record is updated in the access control chain.

3.4 Multi-authority Management

The reliance on a single authority can result in a single point of failure [20]. To mitigate this, the multi-attribute authority was used to manage and maintain attributes. To ensure distribution of attribute authority, the paper employs the Shamir secret sharing (SSS) scheme. SSS is a cryptographic secret sharing scheme that utilizes principles from linear algebra [21]. This is achieved in the following two steps.

(1) Share: A secret value s is divided into n parts, creating a secret sharing scheme. To reconstruct the original secret value s, at least t parts out of the n parts are required, first build a $(t - 1)$-degree polynomial $f(x) = a_0 + a_1 x + \ldots a_{t-1} x^{(t-1)}$. Where $a_0 = s, a_1 \ldots a_{t-1}$ are randomly chosen from Zp. Then, the shares for n participants are generated as $s_i = f(x_i)(i = 1, \ldots, n)$, where $x_i \in Zp$.

(2) Reconstruct: Reconstruction of secret values using LaGrange interpolation. In n shares $s_1 \ldots s_n$, any t shares are selected to compute the s using the Formula (1).

$$s = \Sigma_{i=1}^{t-1} s_i \Pi_{m=0, m \neq j}^{t-1} \frac{x_m}{x_m - x_j} \tag{1}$$

After registering in the system, a user applies for attribute permission from AAi. An attribute permission α is divided into n parts, and a user can apply for at least t parts from different AAi to reconstruct α and obtain the corresponding attribute permission. Contract to calculate the user's attributes Hash value $hi = \text{Hash}\,(uid\,|a_i, attr_i)$, including the uid said user id, α represents the property permission, attr attribute values. And the attribute information $info$ is updated to the access control chain to ensure that user attributes are immutable. The user information is shown in Formula (2).

$$info = \{uid, userAttr < h_1, h_2 \ldots h_m >\} \tag{2}$$

4   Permission-Token and Attributes Based Access Control Model

In this paper, we propose an access control model called PT-ABAC (Permission-Token and Attributes Based Access Control) that merges ABAC and CapBAC. The model is implemented using smart contracts. PT-ABAC expands on the ABAC model by introducing permission token to categorize attributes as static or dynamic within ABAC. Among them, static attributes refer to entity attributes owned by users that are not subject to frequent change, and verification of these is achieved by granting permission tokens to users. Dynamic attributes comprise user-related environmental attributes, such as time, location, and IP address. By validating the user's static attributes beforehand and granting the corresponding permission token, data access efficiency can be enhanced, eliminating the need for repeated validation during multiple, batch accesses to credit data within a short time frame (Fig. 2).

4.1   Permission Token Application

The user first performs attribute authentication off-chain, provides his own attribute to the AAi, and the AAi will generate a zk-SNARK program "Prove" to verify the attribute,

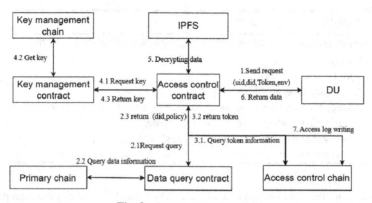

Fig. 2. Access control model

and if it passes the verification, it will generate a zero-knowledge proof for the user. The user applies for the permission token to the Verify contract by providing proof π. The user who obtains the permission token can access the relevant data for many times and in batches within the period. The permission token is shown in Eq. (3).

$$token = \{uid, date, resource < data_1 \ldots data_m >\} \qquad (3)$$

The *uid* is the unique identification of the user, *date* is the duration of the token, and *resource* is the data resource that can be accessed (Fig. 3).

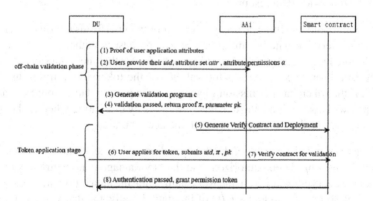

Fig. 3. Request permission token

The user initiates an attribute proof π request to *AAi*, and *AAi* generates a zk-SNARK program named "Prove", which is used to verify the validity of the user attribute and the compliance of the policy, as shown in Algorithm 1. The user inputs parameter *pk*, public input (*policy and* user attribute information *userAttr*), and private input (attribute set *attr*, attribute permission α). If the verification is successful, the *AAi* generates the proof π and provides it to the user. Subsequently, *AAi* deploies the corresponding verification contract "Verify" on the blockchain.

Algorithm 1 Prove program

Input: *pk, uid, private input(attr, a_i), public input (policy, userAttr)*
Output: *result*
1.*for i in 0..len(policy)*
2.   *assert(userAttr[i]==Hash(uid | a_i, Attr[i].value))*
//Verify that the user attributes match the information in the chain
3.   *assert(UserAtt[i].value $\propto$ policy[i].value)*
/*Verify that the user attribute value satisfies the policy */
4.*return true*

The user applies for a permission token to the contract. The Verify contract performs two steps: firstly, it checks the validity of the user's identity to ensure authenticity and eligibility for requested permission; secondly, it verifies the correctness of the proof provided by the user, and if the verification process is successful, the Verify contract grants the user the corresponding permission token.

4.2 Access Control Process

a) Application for access: The requester sends a request message that contains a quaternion consisting of the user ID *uid*, data ID *did*, token ID *tokenId*, and environment attributes *env*. The access control contract first parses the request message to extract relevant information, such as the user's permission token and the policy associated with the requested data resource.

b) Call contract judgement: The access control contract performs a series of verification steps to determine whether the user's access request is authorized or not. First, it verifies the legitimacy of the user's identity to ensure that the requester is a valid and authorized user. Next, it checks the validity of the token by verifying its *date* and check if the token has permission to access the requested data resource. Finally, the contract will also validate the dynamic attributes by evaluating whether the provided environment attributes *env* comply with the access policy.

c) Getting the data: If the result of the policy judgment is true, it indicates that the user's access request has been authorized, and the key management contract is invoked to obtain the decryption key from the KeyChain. Then the data is obtained and decrypted according to the hash address *CID* of the data. Finally, the decrypted data resource will be returned to the requester to update the data access record on the access control chain.

Algorithm 2 Access control algorithm

Input: *uid, tokenid, did, env*
Output: *result*
1.usersigner = Sign(usk)
2.if usersigner != GetUser(uid).signer   / Verify User Signature*/*
3.   return false
*4.token,err = GetToken(uid,tokenid) /*Get the token*/*
5.date = token.date
6.datetime = datetime.Now()
*7.if date.After(datetime)   /*Verify token duration*/*
8   return false
*9.if did lε token.resource   /*Verify access permission*/*
10.   return false
*11.policy = GetData(did).envPolicy   /*Get the data env policy*/*
11.for i :=0 ;i<len(policy);i++{
12.   if env[i].value ∝policy[i].value / Validation attribute*/*
13.return true

---

5 Security Analysis

In this article, the main concern is to ensure the security of access to credit data sharing, with particular emphasis on preventing unauthorized users beyond its authorized access. The goal is to establish a robust and reliable access control mechanism to safeguard credit data from unauthorized or inappropriate access and to protect users' personal privacy. The attribute authority AA is considered a trusted entity. Hence, users attempting to access data beyond their authority can only obtain permission token by forging attribute proofs and tampering with contracts. However, if a user tries to forge their attributes to pass the off-chain "Prove" program and obtain the verification proof π for applying for authority token, they would need to provide false information such as attribute *attr* and attribute permission a_i. . Consequently, the resulting hash value of Hash $(uid \, | a_i, attr_i)$ would be inconsistent with the hash value stored on the blockchain. As hash functions are irreversible, the user would be unable to pass the on-chain verification program, preventing them from successfully applying for permission token. Furthermore, due to the security provided by zk-SNARK, users are unable to pass the verification process of the Verify contract by randomly generating a proof π. The security of zk-SNARK is based on the hardness of the elliptic curve discrete logarithm problem, which prevents users from randomly generating valid proofs. Literature has verified the security properties of zk-SNARK [19].

On the other hand, modifying contracts stored on the blockchain is nearly impossible. The decentralized nature of the blockchain and the consensus algorithms used in the system guarantee that contracts are unchanging and secure. Once a contract is enacted on the blockchain, it becomes part of the distributed ledger, rendering it impossible for any single user to tamper with or modify it. Consensus algorithms usually require a majority of network participants to agree on the legitimacy of transactions and contract updates, enhanced the integrity and security of the contract. Consequently, a user cannot obtain unauthorized access by tampering with a contract, as the contract is held on the blockchain and safeguarded by the underlying consensus mechanism.

6 Experiment

In this paper, the access efficiency of the proposed access control model is evaluated by comparing with other ABAC schemes through simulation experiments. The model presented in this paper combines ABAC and CapBAC and utilizes empowerment token instead of repeatedly determining static attributes within a short timeframe. This method minimizes unnecessary redundant decisions, and the access control process in this paper is implemented based on smart contracts, and the index structure is set to provide query speed, thus optimizing the access efficiency. To assess the performance of the proposed model, the experiments conducted focus on two aspects. First, the time overhead is evaluated under varying cumulative numbers of access requests. Second, the time overhead is tested under batch access scenarios.

Figure 4 illustrates the cumulative data access time overhead for this paper. Compared to the literature [12], the first access time may be longer due to the fact that users need to request permission token through zero-knowledge proof authentication. However, as the number of accesses increases, the total access time of the proposed PT-ABAC model becomes more favorable compared. This improvement can be attributed to the utilization of permission token, which replaces the repeated determination of static attributes for users and effectively reduces data access time.

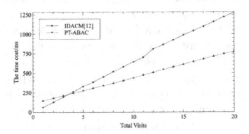

Fig. 4. Cumulative time cost

The Fig. 5 presents the policy determination time for batch access data scenarios. In the permission token, the resource field contains information about the data resources that the user is authorized to access. As a result, the policy determination process only needs to verify whether the user's environment attributes comply with the policy associated with the requested data resource. This streamlined verification process significantly reduces the time required for policy determination.

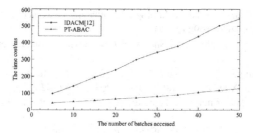

Fig. 5. Batch access time overhead

7 Conclusion

Blockchain technology offers various technical advantages such as security, transparency, anti-tampering, and traceability. These benefits make it a suitable solution for addressing challenges in the field of credit collection. To address these challenges, this paper proposes a personal credit data security sharing scheme, implemented using a multi-chain structure. In the scheme, personal credit data is integrated and stored in IPFS, and the ownership of the data belongs to the information subject. The sharing and utilization of credit data are fully controlled by the information subjects themselves, and access to the credit data is restricted through the implementation of access policies. To ensure strict access control over the data, an efficient data access control model is designed, known as PT-ABAC, which combines ABAC and CapBAC. The model enforces fine-grained access control through the implementation of access policies, preventing unauthorized access by irrelevant individuals. By implementing this solution, the paper achieves secure and efficient sharing of credit data, addressing issues related to data dispersion and low data utilization. Additionally, the scheme establishes clear ownership rights over credit data, effectively protecting the privacy and rights of individuals.

References

1. Ling, D., et al.: Research on the optimization of personal credit investigation index system and information sharing mechanism from the perspective of blockchain. Credit Ref. **40**(05), 1–7 (2022)
2. Ta, L., Li, M.: an analysis of the prospects for application of blockchain technology in internet financial credit. J. Northeastern Univ. (Soc. Sci.) **20**(05), 466–474 (2018)
3. Liu, C., Hou, C.-C.: Challenges of credit reference based on big data technology in China. Mob. Netw. Appl. **27**(1), 47–57 (2021)
4. Tan, H.: Research on improving credit system under the background of consumer privacy protection in internet finance. Times Finance **11**, 4–5 (2020)
5. Zhang, J., et al.: Design and application of a personal credit information sharing platform based on consortium blockchain. J. Inf. Secur. Appl. **55**, 102659 (2020)
6. Ju, C., Zou, J., Fu, X.-K.: Design and application of big data credit reporting platform integrating blockchain technology. Comput. Sci. **45**(S2), 522–526+552 (2018)
7. Ling, C.C., Yang, S., Han, Y.: Research on decentralized model for credit information system. Comput. Technol. Dev. **29**(03), 122–126 (2019)

8. Liang, F., et al.: A survey of key technologies in attribute-based access control scheme. Chin. J. Comput. **40**(07), 1680–1698 (2017)
9. Xin, S., Qing-qi, P., Xue-feng, L.: Survey of block chain. Chin. J. Netw. Inf. Secur. **2**(11), 11–20 (2016)
10. Zhang, Z.W., Wang, G.R., Xu, J.L., Du, X.Y.: Survey on data management in blockchain systems. Ruan Jian Xue Bao/ J. Softw. **31**(9), 2903–2925 (2020). (in Chinese)
11. Qing, C.X., et al.: The principle and core technology of blockchain. Chin. J. Comput. **44**(01), 84–131 (2021)
12. He, H., Yan, A., Chen, Z.: Survey of smart contract technology and application based on blockchain. J. Comput. Res. Dev. **55**(11), 2452–2466 (2018)
13. Zhang, J.B., Zhang, Z.Q., Xu, W.S., Wu, N.: Inter-domain access control model based on blockchain. Ruan Jian Xue Bao/J. Softw. **32**(5), 1547–1564 (2021). (in Chinese)
14. Jian, C., Junpei, N., Jiang, X., et al.: SynergyChain: a multichain-based data-sharing framework with hierarchical access control. IEEE Internet Things J. **9**(16), 14767–14778 (2022)
15. Zia, U., Basit, R., Habib, S., et al.: Towards blockchain-based secure storage and trusted data sharing scheme for IoT environment. IEEE Access **10**, 36978–36994 (2022)
16. Houren, X., et al.: Survey of security analysis for role-based access control. Appl. Res. Comput. **32**(11), 3201–3208 (2015)
17. Liu, Y., et al.: Capability-based IoT access control using blockchain. Digit. Commun. Netw. **7**(4), 463–469 (2021)
18. Song, L., et al.: An access control model for the Internet of Things based on zero-knowledge token and blockchain. EURASIP J. Wirel. Commun. Netw. **2021**(1), 105 (2021)
19. Groth, J.: On the size of pairing-based non-interactive arguments. In: Fischlin, M., Coron, J.-S. (eds.) EUROCRYPT 2016. LNCS, vol. 9666, pp. 305–326. Springer, Heidelberg (2016). https://doi.org/10.1007/978-3-662-49896-5_11
20. Qin, X., et al.: A blockchain-based access control scheme with multiple attribute authorities for secure cloud data sharing. J. Syst. Archit. **112**, 101854 (2021)
21. Hao, G., Meamari, E., Shen, C.-C.: Multi-authority attribute-based access control with smart contract. In: Proceedings of 2019 International Conference on Blockchain Technology, pp. 6–11 (2019)

OR-SPESC: Design of an Advanced Smart Contract Language for Data Ownership

Yuefeng Du[1], Chang Lin[1], Tingting Liu[1], Xiaoguang Li[1(✉)], Wei Wei[2,3], and Shanshan Gao[4]

[1] Liaoning University, Shenyang 110036, China
xgli@lnu.edu.cn
[2] Weihai Ocean Vocational College, Weihai 264315, China
[3] Weihai Ocean Big Data Intelligent Application Engineering Technology Research Center, Weihai 264315, China
[4] Liaoning Province Military Region Data Information Office, Shenyang 110032, China

Abstract. Owing to the open and sharing characteristics, blockchain can be applied for data ownership management in data circulation. The smart contract, as a kernel technique of blockchain, is a program code that can automatically execute the business process of the scene. Smart contracts require programmers with professional coding ability in contract design and implementation. Hence, it is necessary to design an auxiliary tool for non-coding personnel to write smart contracts and manage data ownership as well. To cope with this problem, this paper proposes an advanced smart contract language OR-SPESC for data ownership. OR-SPESC is a contract language similar to natural language including five parts of parties, data assets, deeds, terms, and contract properties. It can select the appropriate function meta-language for a specific scenario to complete the scenario business. Firstly, OR-SPESC makes a formal definition of deed, which describes the ownership between parties and data assets. This paper also proposes three operations on data ownership including creation, destruction, and transfer. Then, taking the trade scenario as an example, the design of OR-SPESC contract rules. Finally, the experiments implement the tracking of OR-SPESC contract conversion and the analysis and verification of the conversion rate (CR) and product rate (PR) of OR-SPESC contract conversion. The research illustrates the effectiveness of OR-SPESC contract design and the efficiency of contract conversion.

Keywords: Data circulation · Ownership · ASCL · Contract conversion

Blockchain, as a new-generation data storage technology, concerns open data sharing, secure and traceability through distributed encryption. This attributes to data assets management and protection under circulation scenarios [1]. Smart contracts, as the core technology of the new blockchain, are digital, automatically executable computer protocols [2], which can also realize the scene functions.

Ownership describes the circulation relationship of owning, holding, using, and operating between subjects and data assets. In the process of data circulation, blockchain needs to use ownership to realize the value attribution and value distribution of data assets according to the functional requirements of the scenario. Currently, owing to the lack of

X. Song et al. (Eds.): APWeb-WAIM Workshops 2023, CCIS 2094, pp. 77–88, 2024.
https://doi.org/10.1007/978-981-97-2991-3_7

relevant research on the use of smart contracts for ownership management, we specify the challenges as follows:

- Existing smart contracts focus on the design of business logic and the realization of processes, and although it is possible to stipulate the ownership in the contract, there is a lack of structures specialized in storing data ownership. When dealing with business functions in some special scenarios, the ownership information is not clear and the functions are relatively limited.
- Traditional smart contract programming methods are not friendly to business people for contract writing. Business people are usually non-programming professionals who only understand the requirements of the scenario and the ownership, and are unable to convert business functions into the code form of smart contracts. Therefore, there is a need to design an auxiliary tool that can support business personnel to perform semi-automatic writing of smart contracts.

Based on this, this paper proposes OR-SPESC, an advanced smart contract language oriented to data ownership, which can realize functional design according to the scenarios and be converted into the target smart contract to complete the data ownership operations. Specifically, the contributions of this paper are as follows:

(1) An advanced smart contract language OR-SPESC for data ownership is proposed. A formal definition of a deed is proposed to store the ownership of data assets. The operations related to the creation, destruction, and transfer of ownership are given.
(2) A conversion rule from OR-SPESC to the target smart contract language (Go language) is proposed. Firstly, an OR-SPESC organizational structure for business scenarios and meta-language for transaction transfer are proposed.
(3) Taking the transaction scenario as an example, we designed an example of OR-SPESC's term rules.
(4) Experiments are conducted to track the implementation of OR-SPESC contract functions and compare the code conversion to verify the effectiveness and efficiency of the OR-SPESC proposed in this paper.

In this paper, Sect. 1 introduces the related work; Sect. 2 proposes a formal definition of the organizational structure and deeds of the advanced smart contract OR-SPESC, and designs the functional meta-language of OR-SPESC and the related operations of the ownership; Sect. 3 takes data transaction scenario as an example of the design of the term rule instance of OR-SPESC; Sect. 4 conducts an experimental analysis and discussion; and Sect. 5 summarizes the full work.

1 Related Work

The research in this paper is about the management problem of data asset ownership and the conversion problem of smart contracts, which belong to the hot issues of research in the field of blockchain applications. Currently, the most closely related work to this research includes the following two aspects.

Domain application of data asset ownership. Data assets have different value attributes and ownership management rules in different domains. Gu Qin *et al.* constructed the data ownership system in data circulation and categorized the ownership

[3]. Wang Di *et al.* proposed a legal contract system architecture with data ownership and explained its deployment and operation [4]. These methods provide theoretical explanations of ownership from the nature and basic theories of ownership in data circulation, but they lack the scenario implementation of ownership using smart contracts.

Advanced smart contract language conversion. The smart contract is essentially a program that can be automatically executed, mainly including script type [5], traditional language type [6], and specialized type [7]. Advanced Smart Contract Language (ASCL) is often regarded as an intermediate auxiliary tool for the conversion from real contracts to the three types of smart contracts. Chen *et al.* proposed an advanced smart language contract SPESC [8], which can support the conversion of contract scenario to Solidity code. Base on SPESC, Zhu *et al.* proposed an asset transaction-oriented TA-SPESC language that binds real assets with ownership information and describes the process of exchanging ownership in a transaction [9]. Arusoaie *et al.* proposed Findle, an advanced smart contract language for the financial field, which realizes the functions of contract fund transfer, logic, and timing expression [10].

Existing ASCLs still lack specialized ownership storage structures and management methods in the process of transaction implementation as well as language conversion. Based on SPESC, we address the design problem of an advanced smart contract language oriented to data ownership.

2   OR-SPESC Language for Ownership

Ownership describes the association relationship between parties and data assets, in order to describe the role and impact of ownership in circulation scenarios, this paper proposes an advanced smart contract language OR-SPESC (Ownership SPESC) oriented to ownership.

2.1   A Subsection Sample

The collection of data circulation scenarios S mainly includes data registration, transaction, sharing, service, hosting, revenue distribution, etc. [3], which are labeled as $S = \{S_1, S_2,...\}$, $S_1 = $ 'Register', $S_2 = $ 'Purchase'. In this paper, we focus on the three ownerships involved in data circulation [9], namely, the right to hold, the right to use, and the right to operate, which can impose restrictions on scenario actions. In different scenarios, parties with specific ownership are able to perform specific actions.

Figure 1 takes the data transaction scenario S_2 as an example, the problem of transferring ownership for data circulation is defined as follows: In S_2, the sender of the transaction, the *sponsors* use the schema method of contracts to send the data asset to the *receptors*, and the deed controlling the ownership is changed from *deed* to *deed'*. The transaction needs to store the *asset* information, the target smart contract, the original *deed*, and the execution result (containing the changed *deed'*) on the chain.

2.2   OR-SPESC Structure

Definition 1 *ORContract$_{S_i}$*. *ORContract$_{S_i}$* is a scenario-specific functionality-based natural language model consisting of five parts: parties, data assets, deeds, terms, and

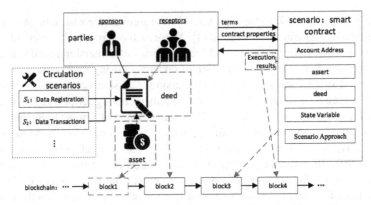

Fig. 1. The sample of the transaction scene with ownership transfer

```
ORContract PurchaseName:Purchase{
party actors:Purchase (sponsors, receptors)
assert assertName {/* details of assert*/}
deed deedID (sponsors, assertName, ownership)
term term1:Purchase {... /* details of term1*/}
term term2:Purchase {...}
...{/* other terms*/}
property purchaseInfo {
          price: double,
          quantity: double
          }
}
```

Fig. 2. The code description of OR-SPESC

contract properties. For scenario S_i, $ORContract_{S_i}$ is defined by Eq. 1:

$$ORContract_{S_i} ::= Cname_{S_i}\{party_{S_i} + asset + deed + terms_{S_i} + property\} \quad (1)$$

where $Cname_{S_i}$ is the contract name. $party_{S_i}$ is the party that distinguishes roles. $terms_{S_i}$ is the term, and the party can execute the specified action in $terms_{S_i}$ according to the scenario. The setting of the roles and the terms and conditions that the roles can execute are determined by scenario S_1. *Asset, deed,* and *property* denote data assets, ownership contracts, and contractual properties, respectively. Among them, *property* is used to represent the necessary information required to execute the scenario functions, such as the version information of the contract, and the pricing and quantity of the transaction in the transaction scenario.

Example. The code description of the OR-SPESC contract in the trading scenario S_2 is given in Fig. 2, where **ORContract, party, assert, deed, term,** and **property** are used as keywords to define the structural information of the contract.

2.3 Parties and Roles

For different scenarios, the slave relationship of the parties in the scenario needs to be specified. For example, in a trafficking relationship for a data transaction, it needs to be specified who is the seller (sponsor) and who is the buyer (receptor).

Definition 2 *party$_{S_i}$.* *party$_{S_i}$* is a collection of contract subjects. Participant *party$_{S_i}$* of the scenario S_i as a variable tuple is defined by Eq. 2:

$$party_{S_i} ::= actorsName : S_i(role_1, ..., role_n) \tag{2}$$

actorsName:S_i denotes the name of the set of all parties in scene S_i. $(role_1,...,role_n)$ is a variable tuple that adjusts the number of roles *role* according to the scenario S_i. The value of n and the meaning of the role indicated by the position of the element in the tuple is determined by the scenario S_i.

Definition 3 *role.* *role* consists of a collection of users as well as the scenario actions of the role, as shown in Eq. 3:

$$role ::= roleName\{users + actions_{S_i}\} \tag{3}$$

roleName is the role's name and *users* are the set of users about role, corresponding to the user's account address. *actions$_{S_i}$* is the action that *role* assigned in the scenario S_i.

2.4 Data Asset

Data assets are designed as a separate data structure that retains only the natural properties of the data itself.

Definition 4 *asset.* *asset* is a data structure that represents the natural attributes of data in Eq. 4:

$$asset ::= assetName\{assetProperty + assetName()\} \tag{4}$$

where *assetName* is the name of the data asset and *assetProperty* is the natural attribute information of the data asset, such as textual content of the data. *AssetName()* is the construction method of the data asset. *AssetName()* contains the operations related to the creation of the ownership, which will be described in detail in Sect. 2.8.

2.5 Deeds

There exists an ownership between data assets and users in the flow of data such as holding, using, and operating, which is expressed as the user's power and obligation to dominate, process, and distribute the data assets. In order to link the ownership between users and data assets, this paper proposes the definition of a deed.

Definition 5 *deed. deed* is a ternary group of users, data asset, and ownership as shown in Eq. 5:

$$deed ::= deedID(Users, asset, ownership) \tag{5}$$

where *deedID* denotes the contract ID and *Users* denotes the set of users associated with the data asset that is the subject of the contract. *Ownership* is *Users* about the ownership possessed by *asset*, including the right to possess, the right to use, and the right to franchise, i.e., *ownership* $\subset$ *{possess, use, franchise}*.

2.6 Terms

A contract can be broken down into multiple terms that perform the functions of a scenario in a specified sequence.

Definition 6 *term$_{S_i}$. term$_{S_i}$* is the statement used to realize a specific action in the scenario S_i as shown in Eq. 6:

$$
\begin{aligned}
term_{S_i} ::= \ &termName : S_i\{role \ (\textbf{can}|\textbf{must}) \ action_{S_i} \\
&(\textbf{when} \ perCondition) \\
&(\textbf{while} \ transactions) \\
&(\textbf{where} \ postCondition)\}
\end{aligned} \tag{6}
$$

The keyword *termName:S_i* is the name of the term in the scene S_i, *role* is the role in *party$_{S_i}$* that completes the term in the scene S_i, *action$_{S_i}$* and is the scene action *roles* that will be automatically assigned and realized in the scene S_i. The keyword **can|must** indicates whether *roles* "can" or "must" execute the term in *action$_{S_i}$*. The keywords **when, while**, and **where** denote pre-conditions, transaction transfers, and export conditions respectively.

Example. Figure 3 gives an example of some of the terms of the transaction scenario. For example, term4: Purchase indicates that the sponsor must transfer the packet data1 to the receptor within 3 days of the receptor's payment.

2.7 Translation Transfer and Expressions

There are two basic transaction forms in the transaction scenario: "swap transaction" and "token transaction", which can be realized using three operations: **deposit, withdraw,** and **trade**. To this end, this paper proposes a transaction transfer meta-language to control the transaction process and an expression meta-language to control the program and timing.

```
term term1:Purchase{
sponsors must assetRegister while deposit data1=
purchaseInfo.price
}
term term2:Purchase{
receptors must pay
when sponsors did assetRegister
while deposit $price=data1
}
term term3:Purchase{
receptors can cancel when before receptors did
pay
}
term term4:Purchase{
sponsors must post when within 3 days after
receptors did pay while trade data1 to receptors
}
```

Fig. 3. The sample of terms

Definition 7 *transactions*. *transactions* are the transfer metalanguage for completing a transaction and contains three operations **deposit, withdraw, trade** in Eq. 7:

$$transactions ::= \textbf{deposit } (\$value \ op) \ assetexp$$
$$|\textbf{withdraw } (\$value|assetexp) \qquad (7)$$
$$|\textbf{trade } assetexp \textbf{ to } actorsName : S_i$$

where **deposit** represents the deposit operation and *$value* represents the deposit of tokens for the completion of the transaction. The transaction party locks *$value* (or the collateral asset *asset'*). *op* is a logical relational comparator that indicates the magnitude of the value *$value* of (or *asset'*) in relation to the value of the target asset *asset*, and $op = \{<, \leq, >, =\}$.**withdraw** indicates a fetch operation on *$value*(or *asset*).**trade** operation represents the transfer operation of the transaction. Where *assetexp* in **trade** represents the target asset expression of the transaction in a token transaction and the collateral asset expression and target asset expression in a swap transaction. *actorsName:S_i* is a party in the transaction.

asset will be constrained and controlled by state conditions such as time, logic, relationship, arithmetic, etc. In this paper, we use the expression approach in the literature [8] to represent the state of data assets.

Definition 8 *assetexp*. *assetexp* is a description of the triggering event associated with the transaction in Eq. 8:

$$assetexp ::= (\textbf{within}) \ time \ (\textbf{before}|\textbf{after})$$
$$role \ \textbf{did } actions_{S_i} \qquad (8)$$

where **within** denotes the time condition that should be met, *time* is the duration variable, **before|after** denotes the event that should be met when the transaction trade is executed, and *roles* **did** $actions_{S_i}$ corresponds to the content of the event.

2.8 Ownership Operations

When executing scenario actions for data circulation, the creation, cancellation, and change of state of the deed will be accompanied. This paper proposes creation, destruction, and transfer operations for ownership management.

Among them, creation and destruction need to be completed in the data registration scenario. The *owner* role uses *assetName()* in Definition 4 for the process of data construction, and calls *bind()* to complete the creation of the ownership, generating the corresponding starting deed. The realization process of the ownership creation *bind()* operation is shown in Fig. 4.

bind(){deed=(role1:S1,assetName,ownership)}

Fig. 4. The bind() operation of ownership creation

owner role through the scene given *logout ()* operation, *logout ()* call *unbind ()* to complete the destruction of the ownership operation, ownership destruction *unbind ()* operation implementation process shown in Fig. 5.

unbind(){deed=(role1:S1,assetName,ownership.del( ))}

Fig. 5. The unbind () operation of ownership destruction

In the trade scenario, the sponsor grants its holding rights to the receptor through the *trade* call *deedGen ()* in Definition 7. The process of implementing the ownership change *deedGen ()* operation is shown in Fig. 6.

deedGen(){deed=(role:S2.update(),assetName,ownership)}

Fig. 6. The deedGen () operation of data ownership

Issues to be clarified: 1) Due to the security consideration of protecting the *deed*, the role requires that the *bind ()*, *unbind ()*, and *deedGen ()* operations cannot be called directly, and they can only be realized by the scenario action. 2) The change of ownership is not to modify the ownership of the original *deed*, but to get the new *deed'* from the original *deed* based on the transfer of the ownership of the transaction action, and package the new *deed'* for uploading to the chain. When verifying the ownership, it is necessary to judge whether it is the latest ownership according to the *timestamp* of the block to which the *deed* belongs.

3 Design of Scene Terms

This subsection will illustrate the design of the OR-SPESC contract rules and the execution process of the transaction by taking the data transaction scenario as an example. For the ownership transfer problem defined in Sect. 2.1, this paper designs a set of term rule sets for data transaction scenarios, supplementing two terms *terms5-terms6* for recovering data assets and deposits on the basis of *terms1-terms4* listed in Fig. 3, as shown in Fig. 7.

Where *term5* means that the receptor must transfer his deposit to the sponsor after the sponsor performs the *post* operation, and *term6* means that the receptor must deposit the deposit to his account within 5 days.

Based on the rule set of *term1–term6*, Fig. 8 depicts the process of OR-SPESC contract executing data transaction, the sender *sponsors* sell the holding right of data asset *data* to *receptors* at the price of *price*, and the transaction process completes the change of *data* belonging to the deed. Step 1: The sponsor lists the data asset for transaction for pricing, and checks the data on chain and the status of the contract, and verifies whether the sponsor has the qualification to trade the data. Step 2: The receptor pays the collateral according to the pricing. Step 3: The sponsor transfers the data asset to the receptor and generates a new deed after the receptor pays the money. Step 4: The receptor transfers the transaction money to the sponsor and recovers the data asset. Step 5: The sponsor obtains the transaction money. Through the above steps, the data transaction is completed.

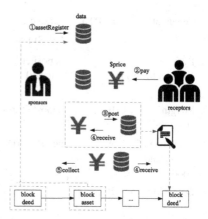

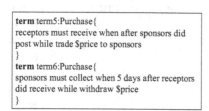

Fig. 7. The supplied terms in the purchase scene

Fig. 8. The trade process of OR-SPESC

4   Experiments

Experiments on contract tracking execution, CR (conversion rate) and PR (product rate) are conducted to verify the effectiveness and efficiency of OR-SPESC.

4.1   Experimental Setting

Execution. We taking the trade scenario as an example, the code tracking of contract execution is shown in Table 1. From the account balances of the parties in Table 1, it can be seen that OR-SPESC, SPESC, and TA-SPESC can all complete the trading operation normally, which reflects the correctness of the advanced smart contract language conversion execution program, and illustrates the effectiveness of the OR-SPESC design. Specifically, the parameter tracking of the contract execution process shows that OR-SPESC can price and collateralize the tokens and data assets of the transaction, and complete the change of ownership by using the deed of *newDeedID* at the completion of the transaction.

Conversion Effects of Contracts. According to the operation of trading scenarios, the experiment was designed with 10 OR-SPESC instances with different trading conditions. Then the results are analyzed based on the average of the number of code bars of the 10 contracts converted into the target contract, and the results are shown in Table 2.

The CR of OR-SPESC is 89.01%, and the missing part requires manual code completion. OR-SPESC requires less code completion because OR-SPESC designs data assets and deeds as separate data structures, which makes this part of the code conversion more complete.

Table 1. The execution of ASCL

implementation process	PARTY	parameters			account balance
		OR-SPESC	SPESC	TA-SPESC	
1. Data collateralization	A	deposite(data) price:10		regRWA + depositeRWA price:10	A:100, B:100
2. Margin collateral	B	deposite($price:10)	$info:price	CDCA	A:100, B:90
3. Exchange of money and goods	A	newDeedID		RWA.update()	A:100, B:90
4. Data collection	B				A:100, B:90
5. Collection of transaction funds	A			revokeRWA	A:110, B:100

From the product rate PR, it can be seen that the conversion ratio of OR-SPESC is 434.04%, which means that one OR-SPESC statement can be equivalent to 4.3 GO language statements, as shown in Fig. 9. On the one hand, OR-SPESC requires additional code overhead for designing and storing data assets and deeds. On the other hand, the GO language itself is more expressive and can be compressed in less space than the solidity language.

Based on the above experimental results, the effectiveness of OR-SPESC in performing contract design and the efficiency of target contract conversion are illustrated by the operation of OR-SPESC and the conversion of the code.

Table 2. The conversion results of ASCL

Contract Type	(ASCL) LLOC	Target Contract Type	autogenLLOC	modified LLOC	total LLOC	CR (%)	PR (%)
OR-SPESC	65.8	GO language	254.2	31.4	285.6	89.01	434.04
SPESC	62.3	solidity language	354.7	67.9	422.6	83.93	678.33
TA-SPESC	57.0	solidity language	378.4	77.8	456.2	82.95	800.35

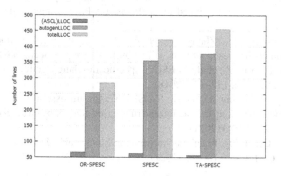

Fig. 9. The conversion performance of contracts

5 Conclusion

In this paper, we propose OR-SPESC, an advanced smart contract language that supports the semi-automated authoring of smart contracts for ownership management by industry personnel. OR-SPESC proposes formal definitions of data assets and deeds, as well as operations related to the creation, destruction, and transfer of ownership. The process and results of OR-SPESC contract design and transformation are also demonstrated through examples and experiments, which illustrate the effectiveness and efficiency of OR-SPESC. In the next step, the conversion process can be standardized and the efficiency of conversion can be improved by designing the PATTERN template for the scenario.

Acknowledgement. This research was supported by the National Key Research and Development Program of China under Grant No. 2021YFF0901004, Research Projects of Liaoning Provincial Department of Education under Grant No. LJKQZ20211023 and No. LJKZ0094, the Projects Shandong Provincial Engineering Center under Grant No. SSCC2030006, the Science and Technology Program Major Project of Liaoning Province of China under Grant No. 2022JH1/10400009, the Natural Science Foundation of Liaoning Province of China under Grant No. 2022-MS-171, the Curriculum Ideology and Politics Model Course Project of Liaoning University.

References

1. Xu, C., Zhang, C., Xu, J., et al.: SlimChain: scaling blockchain transactions through off-chain storage and parallel processing. Proc. VLDB Endow. **14**(11), 2314–2326 (2021)
2. Gupta, S., Hellings, J., Rahnama, S., et al.: Building high throughput permissioned blockchain fabrics. Proc. VLDB Endow. **13**(12), 3441–3444 (2020)
3. Gu, Q., Zhou, T., Zhong, S., et al.: Construction of data tenure system under the two-dimensional perspective of information-data. Big Data **8**(5), 153–169 (2022)
4. Wang, D., Zhu, Y., Chen, E., et al.: Intelligent legal contract and its research progress. J. Eng. Sci. **44**(1), 68–81 (2022)
5. Nathan, S., Govindarajan, C., Saraf, A., et al.: Blockchain meets database: design and implementation of a blockchain relational database. Proc. VLDB Endow. **12**(11), 1539–1552 (2019)
6. Androulaki, E., Manevich, Y., Muralidharan, S., et al.: Hyperledger fabric: a distributed operating system for permissioned blockchains. In: Proceedings of the EuroSys, no. 30, pp. 1–15 (2018)
7. Singh, A., Parizi, R.M., Zhang, Q., et al.: Blockchain smart contracts formalization: approaches and challenges to address vulnerabilities. Comput. Secur. **88**(1), 1–16 (2020)
8. Xiao, H., Qin, B., Yan, Z., et al.: SPESC: a specification language for smart contracts. Proc. IEEE Comput. Soc. (1), 132–137 (2018)
9. Zhu, Y., Song, W., Wang, D., et al.: TA-SPESC: toward asset-driven smart contract language supporting ownership transaction and rule-based generation on blockchain. IEEE Trans. Reliab. **70**(3), 1255–1270 (2021)
10. Arusoaie, A.: Certifying findel derivatives for blockchain. J. Logical Algebr. Methods Program. **121**(6), 100665–100673 (2021)

Author Index

X. Song et al. (Eds.): APWeb-WAIM Workshops 2023, CCIS 2094, p. 89, 2024.
https://doi.org/10.1007/978-981-97-2991-3

Printed in the United States
by Baker & Taylor Publisher Services